Lóegaire Humphrey (Ed.)

Environmental Modeling Center

Lóegaire Humphrey (Ed.)

Environmental Modeling Center

Data assimilation, Climatology, Meteorology

Claud Press

Imprint

Publisher:
Claud Press is a trademark of
International Book Market Service Ltd., 17 Rue Meldrum, Beau Bassin, 1713-01 Mauritius
Email: info@bookmarketservice.com
Website: www.bookmarketservice.com

Published in 2011

Printed in: U.S.A., U.K., Germany. This book was not produced in Mauritius.

ISBN: 978-613-7-28668-5

Contents

Articles

References

Environmental Modeling Center

The **Environmental Modeling Center** (EMC), improves numerical weather, marine and climate predictions at the National Centers for Environmental Prediction (NCEP), through a broad program of research in data assimilation and modeling. In support of the NCEP operational forecasting mission, the EMC develops, improves and monitors data assimilation systems and models of the atmosphere, ocean and coupled system, using advanced methods developed internally as well as cooperatively with scientists from Universities, NOAA Laboratories and other government agencies, and the international scientific community.

Branches

- Global Climate & Weather Modeling
- Mesoscale Modeling
- Marine Modeling and Analysis Branch

External links

- http://www.emc.ncep.noaa.gov/

Data assimilation

Applications of **data assimilation** arise in many fields of geosciences, perhaps most importantly in weather forecasting and hydrology. Data assimilation proceeds by *analysis cycles*. In each analysis cycle, observations of the current (and possibly, past) state of a system are combined with the results from a numerical weather prediction model (the *forecast*) to produce an *analysis*, which is considered as 'the best' estimate of the current state of the system. This is called the *analysis step*. Essentially, the analysis step tries to balance the uncertainty in the data and in the forecast. The model is then advanced in time and its result becomes the forecast in the next analysis cycle.

Data assimilation as statistical estimation

In data assimilation applications, the analysis and forecasts are best thought of as probability distributions. The analysis step is an application of the Bayes theorem and the overall assimilation procedure is an example of Recursive Bayesian estimation. However, the probabilistic analysis is usually simplified to a computationally feasible form. Advancing the probability distribution in time would be done exactly in the general case by the Fokker-Planck equation, but that is unrealistically expensive, so various approximations operating on simplified representations of the probability distributions are used instead. If the probability distributions are normal, they can be represented by their mean and covariance, which gives rise to the Kalman filter. However it is not feasible to maintain the covariance because of the large number of degrees of freedom in the state, so various approximations are used instead.

Many methods represent the probability distributions only by the mean and impute some covariance instead. In the basic form, such analysis step is known as optimal statistical interpolation. Adjusting the initial value of the mathematical model instead of changing the state directly at the analysis time is the essence of the variational methods, 3DVAR and 4DVAR. Nudging, also known as Newtonian relaxation or 4DDA, is essentially the same as proceeding in continuous time rather than in discrete analysis cycles (the Kalman-Bucy filter), again with imputing simplified covariance.

Ensemble Kalman filters represent the probability distribution by an ensemble of simulations, and the covariance is approximated by sample covariance.

In addition to weather forecasting, other uses of DA include trajectory estimation for the Apollo program, GPS, and atmospheric chemistry.

Weather forecasting applications

Data assimilation is a concept encompassing any method for combining observations of variables like temperature, and atmospheric pressure into numerical models as the ones used to predict weather.

In weather forecasting there are 2 main types of data assimilation: 3 dimensional (3DDA) and 4 dimensional (4DDA). In 3DDA only those observations available at the time of analysis are used. In 4DDA the future observations are included (thus, time dimension added).

History of Data Assimilation in Weather forecasting

The first data assimilation methods were called the "objective analyses" (e.g., Cressman algorithm). This was in contrast to the "subjective analyses", when (in past practices) numerical weather predictions (NWP) forecasts were adjusted by meteorologists using their operational expertise. The objective methods used simple interpolation approaches, and thus were 3DDA methods.

The similar 4DDA methods, called "nudging" also exist (e.g. in MM5 NWP model). They are based on the simple idea of Newtonian relaxation (the 2nd axiom of Newton). The idea is to add in the right part of dynamical equations of the model a term that is proportional to the difference of the calculated meteorological variable and the observed value. This term that has a negative sign "keeps" the calculated state vector closer to the observations. Nudging can be interpreted as a variant of the Kalman-Bucy filter (a continuous time version of the Kalman filter) with the gain matrix prescribed rather than obtained from covariances.

The breakthrough in the field of data assimilation was achieved by L. Gandin (1963) who introduced the "statistical interpolation" (or "optimal interpolation") method. His work developed the previous ideas of Kolmogorov. That method is a 3DDA method and is a type of regression analyses, which utilizes information about the spatial distributions of covariance functions of the errors of the "first guess" field (previous forecast) and "true field". These functions are never known. However, the different approximations were assumed.

In fact the optimal interpolation algorithm is the reduced version of the Kalman filtering (KF) algorithm, when the covariance matrices are not calculated from the dynamical equations, but are pre-determined in advance.

Attempts to introduce the KF algorithms as a 4DDA tool for NWP models came later. However, this was (and remains) a very difficult task, since the full version of KF algorithm requires solution of the enormous number of additional equations (~N*N~10**12, where N=Nx*Ny*Nz is the size of the state vector, Nx~100, Ny~100, Nz~100 - the dimensions of the computational grid). To overcome that difficulty the special kind of KF algorithms (approximate or suboptimal KF's) for NWP models were developed. These include, e.g., the Ensemble Kalman filter and the Reduced-Rank Kalman filters (RRSQRT) (e.g., Todling and Cohn, 1994).

Another significant advance in the development of the 4DDA methods was utilizing the optimal control theory (variational approach) in the works of Le Dimet and Talagrand (1986), based on the previous works of G. Marchuk, who was the first to apply that theory in the environmental modeling. The significant advantage of the variational approaches is that the meteorological fields satisfy the dynamical equations of the NWP model and at the same time they minimize the functional, characterizing their difference from observations. Thus, the problem of constrained minimization is solved. The 3DDA variational methods were developed for the first time by Sasaki (1958).

As was shown by Lorenc (1986), all the above-mentioned 4DDA methods are in some limit equivalent, i.e. under some assumptions they minimize the same cost function. However, in practical applications these assumptions are never fulfilled, the different methods perform differently and generally it is not clear what approach (Kalman filtering or variational) is better. The fundamental questions also arise in application of the advanced DA techniques such as convergence of the computational method to the global minimum of the functional to be minimised. For instance, cost function or the set in which the solution is sought can be not convex.

Future Development in NWP

The rapid development of the various data assimilation methods for NWP models is connected with the two main points in the field of numerical weather prediction:

1. Utilizing the observations currently seems to be the most promising chance to improve the quality of the forecasts at the different spatial scales (from the planetary scale to the local city, or even street scale) and time scales.
2. The number of different kinds of available observations (sodars, radars, satellite) is rapidly growing.

The question is: can the principal limit of the predictability of weather forecast models be overcome (and to what extent) with the help of data assimilation?

Cost function

The process of creating the analysis in data assimilation often involves minimization of a "cost function." A typical cost function would be the sum of the squared deviations of the analysis values from the observations weighted by the accuracy of the observations, plus the sum of the squared deviations of the forecast fields and the analyzed fields weighted by the accuracy of the forecast. This has the effect of making sure that the analysis does not drift too far away from observations and forecasts that are known to usually be reliable.

1. 3D-Var

where denotes the background error covariance, the observational error covariance.

2. 4D-var

provided that is linear operator (matrix).

Other applications of Data Assimilation

Data assimilation methods are currently also used in other environmental forecasting problems, e.g. in hydrological forecasting. Basically, the same types of data assimilation methods as those described above are in use there. An example of chemical data assimilation using Autochem can be found at CDACentral [1].

Given the abundance of spacecraft data for other planets in the Solar System, data assimilation is now also applied beyond the Earth to obtain re-analyses of the atmospheric state of extra-terrestrial planets. Mars is the first extra-terrestrial planet which data assimilation has been applied to, so far. Available spacecraft data include, in particular, retrievals of temperature and dust/water ice optical ticknesses from the Thermal Emission Spectrometer onboard NASA's Mars Global Surveyor and the Mars Climate Sounder onboard NASA's Mars Reconnaissance Orbiter. Two methods of data assimilation have been applied to these datasets: an Analysis Correction scheme [2] and two Ensemble Kalman Filter schemes [3] [4] , both using a global circulation model of the martian atmosphere as forward model. The Mars Analysis Correction Data Assimilation (MACDA) dataset is publicly available from the British Atmospheric Data Centre [5] .

Data assimilation is a part of the challenge for every forecasting problem.

Dealing with biased data is a serious challenge in data assimilation. Further development of methods to deal with biases will be of particular use. If there are several instruments observing the same variable then intercomparing them using probability distribution functions can be instructive. Such an analysis is available on line at PDFCentral [6] designed for the validation of observations from the NASA Aura satellite.

References

- R. Daley, Atmospheric data analysis, Cambridge University Press, 1991.
- MM5 community model homepage [7]
- ECMWF Data Assimilation Lecture notes [8]
- Ide, K., P. Courtier, M. Ghil, and A. C. Lorenc (1997) Unified Notation for Data Assimilation: Operational, Sequential and Variational [9] Journal of the Meteorologcial Society of Japan, vol. 75, No. 1B, pp. 181–189
- COMET module "Understanding Data Assimilation" [10]
- Geir Evensen, Data Assimilation. The Ensemble Kalman Filter. Springer, 2007
- John M. LEWIS; S. Lakshmivarahan, Sudarshan Dhall, "Dynamic Data Assimilation : A Least Squares Approach", Encyclopedia of Mathematics and its Applications 104, Cambridge University Press, 2006 (ISBN-13 978-0-521-85155-8 Hardback)

Notes

[1] http://www.cdacentral.info/
[2] http://www.atm.ox.ac.uk/group/gpfd/research.html#marsgcm
[3] http://www.eps.jhu.edu/~mjhoffman/pages/research.html
[4] http://www.marsclimatecenter.com
[5] http://badc.nerc.ac.uk/home/
[6] http://www.pdfcentral.info/
[7] http://www.mmm.ucar.edu/mm5/
[8] http://www.ecmwf.int/newsevents/training/lecture_notes/LN_DA.html
[9] http://www.atmos.ucla.edu/~kayo/data/publication/ide_etal_jmsj97.pdf
[10] http://www.meted.ucar.edu/nwp/pcu1/ic6/frameset.htm

External links

Examples of how variational assimilation is implemented weather forecasting at:

- ECMWF http://www.ecmwf.int/research/ifsdocs/ASSIMILATION/Chap1_Overview2.html
- the Met Office http://www.metoffice.gov.uk/science/creating/first_steps/data_assim.html?zoneid=79046

Other examples of assimilation:

- CDACentral (an example analysis from Chemical Data Assimilation) (http://www.CDAcentral.info/)
- PDFCentral (using PDFs to examine biases and representativeness) (http://www.PDFcentral.info/)
- OpenDA – Open Source Data Assimilation package (http://www.openda.org/joomla/index.php)

National Centers for Environmental Prediction

The United States **National Centers for Environmental Prediction** (NCEP) delivers national and global weather, water, climate and space weather guidance, forecasts, warnings and analyses to its Partners and External User Communities. These products and services are based on a service-science legacy and respond to user needs to protect life and property, enhance the nation's economy and support the nation's growing need for environmental information. It is part of the National Weather Service.

There are nine centers:

1. Aviation Weather Center [1] provides aviation warnings and forecasts of hazardous flight conditions at all levels within domestic and international air space.
2. Climate Prediction Center monitors and forecasts short-term climate fluctuations and provides information on the effects climate patterns can have on the nation.
3. Environmental Modeling Center develops and improves numerical weather, climate, hydrological and ocean prediction through a broad program in partnership with the research community.
4. Hydrometeorological Prediction Center provides nationwide analysis and forecast guidance products out through seven days.
5. NCEP Central Operations [2] sustains and executes the operational suite of numerical analyses and forecast models and prepares NCEP products for dissemination.
6. Ocean Prediction Center [3] issues weather warnings and forecasts out to five days for the Atlantic and Pacific Oceans north of 30 degrees North.
7. Space Weather Prediction Center provides space weather alerts and warnings for disturbances that can affect people and equipment working in space and on earth.
8. Storm Prediction Center provides tornado and severe weather watches for the contiguous United States along with a suite of hazardous weather forecasts.
9. The National Hurricane Center provides forecasts of the movement and strength of tropical weather systems and issues watches and warnings for the North Atlantic and the Eastern Pacific ocean.

External links

- NCEP Homepage [4]
- NOAA Center for Weather and Climate Prediction [5]
- Operational Significant Event Imagery [6] (the OSEI 'Image of the Day' is always located here [7]).
- NCEP on www.top500.org [8]

References

[1] http://www.aviationweather.gov
[2] http://www.nco.ncep.noaa.gov/
[3] http://www.opc.ncep.noaa.gov/
[4] http://www.ncep.noaa.gov/
[5] http://www.ncep.noaa.gov/news/ncwcp/latest.shtml
[6] http://www.osei.noaa.gov/
[7] http://www.osei.noaa.gov/iod.html
[8] http://www.top500.org/site/history/1220

National Oceanic and Atmospheric Administration

National Oceanic and Atmospheric Administration	
NOAA	
Agency overview	
Formed	October 3, 1970
Jurisdiction	Federal government of the United States
Headquarters	Silver Spring, MD
Annual budget	US$4.5 billion (2009) US$4.9 billion (est. 2010) US$5.6 billion (est. 2011)
Agency executive	Jane Lubchenco, Administrator
Parent agency	Department of Commerce
Website	
noaa.gov [1]	

The **National Oceanic and Atmospheric Administration (NOAA)**, pronounced /ˈno(ʊ).ə/, like "noah", is a scientific agency within the United States Department of Commerce focused on the conditions of the oceans and the atmosphere. NOAA warns of dangerous weather, charts seas and skies, guides the use and protection of ocean and coastal resources, and conducts research to improve understanding and stewardship of the environment. In addition to its civilian employees, NOAA research and operations are supported by 300 uniformed service members who make up the NOAA Commissioned Officer Corps. The current Under Secretary of Commerce for Oceans and Atmosphere at the Department of Commerce, and the agency's Administrator, is Dr. Jane Lubchenco, nominated by President Barack Obama and confirmed by the United States Senate on March 19, 2009.[2]

Vision, mission, and goals

NOAA's *strategic vision* is "an informed society that uses a comprehensive understanding of the role of the oceans, coasts, and atmosphere in the global ecosystem to make the best social and economic decisions."

National Weather Service meteorologists preparing a forecast, early 20th century

The agency's mission is "to understand and predict changes in the Earth's environment and conserve and manage coastal and marine resources to meet our nation's economic, social, and environmental needs."

In support of its vision and mission, NOAA has four goals to guide its suite of operations. Each goal corresponds to activities focusing on ecosystems, climate, weather and water, and commerce and transportation. Specifically, NOAA operates to:

- Ensure the sustainable use of resources and balance competing uses of coastal and marine ecosystems, recognizing both their human and natural components.
- Understand changes in climate, including global climate change and the El Niño phenomenon, to ensure that Americans can plan and respond properly.
- Provide data and forecasts for weather and water cycle events, including storms, droughts, and floods.
- Provide weather, climate, and ecosystem information to make sure individual and commercial transportation is safe, efficient, and environmentally sound.

Purpose and function

NOAA plays several specific roles in society, the benefits of which extend beyond the US economy and into the larger global community:

Two NOAA WP-3D Orions.

- *A Supplier of Environmental Information Products.* NOAA supplies information to its customers and partners pertaining to the state of the oceans and the atmosphere. This is clearly manifest in the production of weather warnings and forecasts through the National Weather Service, but NOAA's information products extend to climate, ecosystems, and commerce as well.
- *A Provider of Environmental Stewardship Services.* NOAA is also the steward of U.S. coastal and marine environments. In coordination with federal, state, local, tribal, and international authorities, NOAA manages the use of these environments, regulating fisheries and marine sanctuaries as well as protecting threatened and endangered marine species.
- *A Leader in Applied Scientific Research.* NOAA is intended to be a source of accurate and objective scientific information in the four particular areas of national and global importance identified above: ecosystems, climate, weather and water, and commerce and transportation.

Recognizing that it is essential that we understand the challenges that we face as part of the Earth system in order to create appropriate solutions, NOAA conducts an end-to-end sequence of activities, beginning with scientific discovery and resulting in a number of critical environmental services and products. The five "fundamental activities" are:

- Monitoring and observing Earth systems with instruments and data collection networks.
- Understanding and describing Earth systems through research and analysis of that data.
- Assessing and predicting the changes of these systems over time.
- Engaging, advising, and informing the public and partner organizations with important information.
- Managing resources for the betterment of society, economy and environment.

History and organizational structure

NOAA was formed on October 3, 1970, after Richard Nixon proposed creating a new department to serve a national need "... for better protection of life and property from natural hazards ... for a better understanding of the total environment ... [and] for exploration and development leading to the intelligent use of our marine resources ..." NOAA formed a conglomeration of three existing agencies that were among the oldest in the federal government. They were the United States Coast and Geodetic Survey, formed in 1807; the Weather Bureau, formed in 1870; and the Bureau of Commercial Fisheries, formed in 1871. NOAA was established within the Department of Commerce via the Reorganization Plan No. 4 of 1970. With its ties to the United States Coast and Geodetic Survey, NOAA celebrated 200 years of service in 2007.

Seal of the NOAA Commissioned Corps

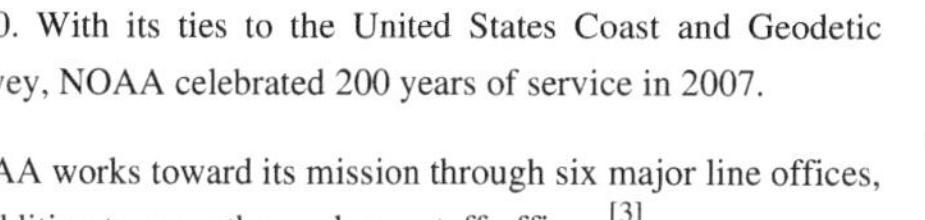

NOAA works toward its mission through six major line offices, in addition to more than a dozen staff offices:[3]

Line Offices

- The National Environmental Satellite, Data and Information Service (NESDIS)
- The National Marine Fisheries Service (NMFS)
- The National Ocean Service (NOS)
- The National Weather Service (NWS)
- Office of Oceanic and Atmospheric Research (OAR)
- Office of Program Planning and Integration (PPI)

Staff Offices

- Office of the Federal Coordinator for Meteorology
- Office of Marine and Aviation Operations
- NOAA Central Library

NOAA Corps

NOAA research and operational activities are supported by a uniformed service, the NOAA Corps. They are a commissioned officer corps of men and women who operate NOAA ships and aircraft, and serve in scientific and administrative posts.

National Weather Service (NWS)

The National Weather Service (NWS) is tasked with providing "weather, hydrologic, and climate forecasts and warnings for the United States, its territories, adjacent waters and ocean areas, for the protection of life and property and the enhancement of the national economy." This is done through a collection of national and regional centers, and more than 120 local weather forecast offices (WFOs). They are charged with issuing weather forecasts, advisories, watches, and warnings on a daily basis. They issue more than 734,000 weather and 850,000 river forecasts, and more than 45,000 severe weather warnings annually. NOAA data is also relevant to the issues of global warming and ozone depletion. The NWS operates NEXRAD, a nationwide network of Doppler weather radars which can detect precipitation and their velocities. Many of their products are broadcast on NOAA Weather Radio, a network of radio transmitters that broadcasts weather forecasts, severe weather statements, watches and warnings 24 hours a day.

Seal of the National Weather Service

National Ocean Service (NOS)

The National Ocean Service (NOS), part of the National Oceanic and Atmospheric Administration within the U.S. Department of Commerce, is focused on ensuring that ocean and coastal areas are safe, healthy, and productive. NOS scientists, natural resource managers, and specialists serve America by ensuring safe and efficient marine transportation, promoting innovative solutions to protect coastal communities, and conserving marine and coastal places.

The National Ocean Service is composed of program offices, programs, and staff offices:

Program Offices

- Center for Operational Oceanographic Products and Services (CO-OPS) [4]
- NOAA Coastal Services Center (CSC)
- National Centers for Coastal Ocean Science (NCCOS)
- Office of Coast Survey (OCS)
- Office of National Geodetic Survey (NGS)
- Office of National Marine Sanctuaries (ONMS)
- Office of Ocean and Coastal Resource Management (OCRM) [5]
- Office of Response and Restoration (OR&R) [6]

Programs

- NOAA Integrated Ocean Observing System (IOOS) Program

Staff Offices

- International Program Office (IPO)
- Management and Budget Office (M&B)

National Environmental Satellite, Data, and Information Service (NESDIS)

The **National Environmental Satellite, Data, and Information Service** (NESDIS) was created by NOAA to operate and manage the United States environmental satellite programs, and manage the data gathered by the NWS and other government agencies and departments. Data collected by the NWS, U.S. Navy, U.S. Air Force, the Federal Aviation Administration, and meteorological services around the world, are housed at the National Climatic Data Center in Asheville, North Carolina. NESDIS also operates the National Geophysical Data Center (NGDC) in Boulder, Colorado, the National Oceanographic Data Center (NODC) in Silver Spring, Maryland, the National Snow and Ice Data Center (NSIDC) and the National Coastal Data Development Center (NCDDC) which are used internationally by environmental scientists.

NOAA engineer at work

NESDIS also runs the:

- Office of Systems Development (OSD) [7]
- Office of Satellite Operations (OSO) [8]
- Office of Satellite Data Processing & Distribution (OSDPD) [9]
- Satellite Applications and Research (STAR) formerly the Office of Research & Applications
- Joint Polar Satellite System Program Office [10]
- GOES-R Program Office
- International & Interagency Affairs Office
- Office of Space Commercialization

The service operates and manages many geosynchronous satellites and polar orbiting satellites. In 1960 TIROS-1, NOAA's first owned and operated geostationary satellite was launched. In 1983 NOAA assumed operational responsibility for LANDSAT satellite system. In 1984 the Tropical Ocean-Global Atmosphere program (TOGA) program began.

In 1977 the Pacific Marine Environmental Laboratory (PMEL) deployed the first successful moored equatorial current meter - the beginning of the Tropical Atmosphere/Ocean (TAO) array. In 1979 NOAA's first polar-orbiting environmental satellite was launched.

Current operational satellites include: NOAA-15, NOAA-16, NOAA-17, NOAA-18 and NOAA-19 (launched 2/6/2009).

National Marine Fisheries Service (NMFS)

Fisheries, which was initiated in 1871 to protect, study, manage and restore fish. The NMFS has a marine fisheries research lab in Woods Hole, Massachusetts and is home to one of NOAA's five fisheries science centers.

Its law enforcement agency is the National Oceanic and Atmospheric Administration Fisheries Office for Law Enforcement based in Silver Spring, Maryland.

Office of Oceanic and Atmospheric Research (OAR)

NOAA's research, conducted through the Office of Oceanic and Atmospheric Research (OAR), is the driving force behind NOAA environmental products and services that protect life and property and promote economic growth. Research, conducted in OAR laboratories and by extramural programs, focuses on enhancing our understanding of environmental phenomena such as tornadoes, hurricanes, climate variability, solar flares, changes in the ozone, air pollution transport and dispersion,[11] [12] El Niño/La Niña events, fisheries productivity, ocean currents, deep sea thermal vents, and coastal ecosystem health. NOAA research also develops innovative technologies and observing systems.

The NOAA Research network consists of 7 internal research laboratories, extramural research at 30 Sea Grant university and research programs, six undersea research centers, a research grants program through the Climate Program Office, and 13 cooperative institutes with academia. Through NOAA and its academic partners, thousands of scientists, engineers, technicians, and graduate students participate in furthering our knowledge of natural phenomena that affect the lives of us all.

The Air Resources Laboratory (ARL) is one of the laboratories in the Office of Oceanic and Atmospheric Research. It studies processes and develops models relating to climate and air quality, including the transport, dispersion, transformation and removal of pollutants from the ambient atmosphere. The emphasis of the ARL's work is on data interpretation, technology development and transfer. The specific goal of ARL research is to improve and eventually to institutionalize prediction of trends, dispersion of air pollutant plumes, air quality, atmospheric deposition, and related variables.

National Geodetic Survey

The National Geodetic Survey is the primary surveying organization in the United States.

National Integrated Drought Information System

NOAA is the lead federal agency for the National Integrated Drought Information System (NIDIS).

Program Planning and Integration (PPI)

The Office of Program Planning and Integration was established in June 2002 as the focus for a new corporate management culture at NOAA. PPI was created to address the needs to:

- Foster strategic management among NOAA Line and Staff Offices, Goal Teams, Programs, and Councils,
- Support planning activities through greater opportunities for active participation of employees, stakeholders, and partners,
- Build decision support systems based on the goals and outcomes set in NOAA's strategic plan, and
- Guide managers and employees on program and performance management, the National Environmental Policy Act, and socioeconomic analysis.

Intergovernmental Panel on Climate Change

Since 2001 the organization has hosted the senior staff and recent chair, Susan Solomon, of the Intergovernmental Panel on Climate Change's working group on climate science.[13]

Flag

The National Oceanic and Atmospheric Administration flag, flown as a distinguishing mark by all commissioned NOAA ships.

The NOAA flag is a modification of the flag of one of its predecessor organizations, the United States Coast and Geodetic Survey. The Coast and Geodetic Survey's flag, authorized in 1899 and in use until 1970, was blue, with a white circle centered in it and a red triangle centered within the circle. It symbolized the use of triangulation in surveying, and was flown by ships of the Survey.

When NOAA was established in 1970 and the Coast and Geodetic Survey's assets became a part of NOAA, NOAA based its own flag on that of the Coast and Geodetic Survey. The NOAA flag is in essence the Coast and Geodetic Survey flag, with the NOAA logo—a circle divided by the silhouette of a seabird into an upper dark blue and a lower light blue section, but with the "NOAA" legend omitted—centered within the red triangle. NOAA ships in commission display the NOAA flag; those with only one mast fly it immediately beneath the ship's commissioning pennant or the personal flag of a civilian official or flag officer if one is aboard the ship, while multimasted vessels fly it at the masthead of the forwardmost mast.[14] NOAA ships fly the same ensign as United States Navy ships but fly the NOAA flag as a distinguishing mark to differentiate themselves from Navy ships.

See also

- Robert Ballard
- Center for Environmental Technology (CET)
- Endangered Species Act
- Federation of Earth Science Information Partners (ESIP Federation)
- Marine Mammal Protection Act
- National Oceanic and Atmospheric Administration Commissioned Corps
- Minerals Management Service
- NOAA's Environmental Real-time Observation Network
- NOAA's Virtual World Program
- Volcanic Ash Advisory Center
- Weather Modification Operations and Research Board
- Office of Naval Research
- United States Naval Research Laboratory
- University-National Oceanographic Laboratory System
- List of auxiliaries of the United States Navy
- Center for Ocean Solutions [15]

References

[1] http://www.noaa.gov
[2] U.S. Senate: Nominations Confirmed. (http://www.senate.gov/pagelayout/legislative/one_item_and_teasers/nom_confc.htm) Accessed March 21st, 2009.
[3] NOAA Organizations (http://www.noaa.gov/organizations.html)
[4] http://tidesandcurrents.noaa.gov/
[5] http://coastalmanagement.noaa.gov/
[6] http://response.restoration.noaa.gov/
[7] http://www.osd.noaa.gov
[8] http://www.oso.noaa.gov
[9] http://www.osdpd.noaa.gov/ml/index.html
[10] http://www.nesdis.noaa.gov/jpss/
[11] Turner, D.B. (1994). *Workbook of atmospheric dispersion estimates: an introduction to dispersion modeling* (2nd Edition ed.). CRC Press. ISBN 1-56670-023-X. CRCpress.com (http://www.crcpress.com/shopping_cart/products/product_detail.asp?sku=L1023&parent_id=&pc=)
[12] Beychok, M.R. (2005). *Fundamentals Of Stack Gas Dispersion* (4th Edition ed.). author-published. ISBN 0-9644588-0-2. www.air-dispersion.com (http://www.air-dispersion.com)
[13] Pearce, Fred, *The Climate Files: The Battle for the Truth about Global Warming*, (2010) Guardian Books, ISBN: 978-0-85265-229-9, p. XVIII.
[14] Sea Flags: National Oceanic and Atmospheric Administration (http://mysite.verizon.net/vzeohzt4/Seaflags/noaa/noaa.html)
[15] http://www.centerforoceansolutions.org

External links

- NOAA home page (http://www.noaa.gov)

Marine Modeling and Analysis Branch

The United States **Marine Modeling and Analysis Branch** (MMAB) is part of the Environmental Modeling Center, which is responsible for the development of improved numerical weather and marine prediction modeling systems within NCEP/NWS. It provides analysis and real-time forecast guidance (1–16 days) on marine meteorological, oceanographic, and cryospheric parameters over the global oceans and coastal areas of the US.

Products include:

- Ocean Waves
- Sea ice [1]
- Marine Meteorology
  - Marine Winds - Satellite Remote Sensing
  - Coastal Ocean Visibility
  - Open Ocean Visibility
  - Vessel Icing
- Sea surface temperature
- Real-Time Ocean Forecast System [2]
- Automated Gulf Stream

External links

- http://polar.ncep.noaa.gov/

References

[1] http://polar.ncep.noaa.gov/seaice/
[2] http://polar.ncep.noaa.gov/ofs/

Climatology

Climatology (from Greek κλίμα, *klima*, "place, zone"; and -λογία, *-logia*) is the study of climate, scientifically defined as weather conditions averaged over a period of time,[1] and is a branch of the atmospheric sciences. Basic knowledge of climate can be used within shorter term weather forecasting using analog techniques such as the El Niño – Southern Oscillation (ENSO), the Madden-Julian Oscillation (MJO), the North Atlantic Oscillation (NAO), the Northern Annualar Mode (NAM), the Arctic oscillation (AO), the Northern Pacific (NP) Index, the Pacific Decadal Oscillation (PDO), and the Interdecadal Pacific Oscillation (IPO). Climate models are used for a variety of purposes from study of the dynamics of the weather and climate system to projections of future climate.

History

The earliest person to hypothesize the concept of climate change may have been the medieval Chinese scientist Shen Kuo (1031–95). Shen Kuo theorized that climates naturally shifted over an enormous span of time, after observing petrified bamboos found underground near Yanzhou (modern day Yan'an, Shaanxi province), a dry climate area unsuitable for the growth of bamboo.

Early climate researchers include Edmund Halley, who published a map of the trade winds in 1686, after a voyage to the southern hemisphere. Benjamin Franklin, in the 18th century, was the first to map the course of the Gulf Stream for use in sending mail overseas from the United States to Europe. Francis Galton invented the term *anticyclone*.[2] Helmut Landsberg led to statistical analysis being used in climatology, which led to its evolution into a physical sciences.

Different approaches

Climatology is approached in a variety of ways. **Paleoclimatology** seeks to reconstruct past climates by examining records such as ice cores and tree rings (dendroclimatology). **Paleotempestology** uses these same records to help determine hurricane frequency over millennia. The study of contemporary climates incorporates meteorological data accumulated over many years, such as records of rainfall, temperature and atmospheric composition. Knowledge of the atmosphere and its dynamics is also embodied in models, either statistical or mathematical, which help by integrating different observations and testing how they fit together. Modeling is used for understanding past, present and potential future climates. **Historical climatology** is the study of climate as related to human history and thus focuses only on the last few thousand years.

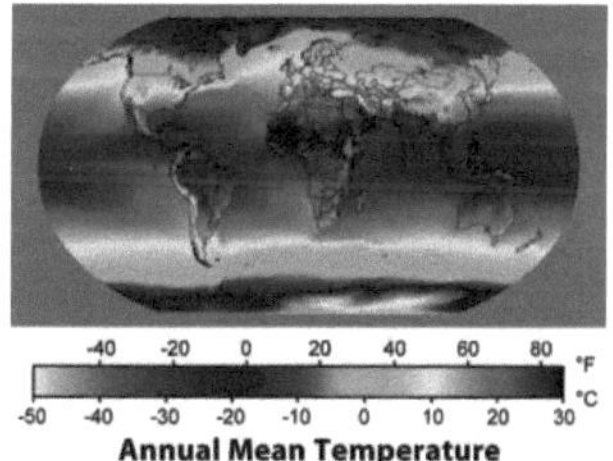

Map of the average temperature over 30 years. Data sets formed from the long-term average of historical weather parameters are sometimes called a "climatology".

Climate research is made difficult by the large scale, long time periods, and complex processes which govern climate. Climate is governed by physical laws which can be expressed as differential equations. These equations are coupled and nonlinear, so that approximate solutions are obtained by using numerical methods to create global climate models. Climate is sometimes modeled as a stochastic process but this is generally accepted as an approximation to processes that are otherwise too complicated to analyze.

Indices

Scientists use climate indices based on several climate patterns (known as modes of variability) in their attempt to characterize and understand the various climate mechanisms that culminate in our daily weather. Much in the way the Dow Jones Industrial Average, which is based on the stock prices of 30 companies, is used to represent the fluctuations in the stock market as a whole, climate indices are used to represent the essential elements of climate. Climate indices are generally devised with the twin objectives of simplicity and completeness, and each index typically represents the status and timing of the climate factor it represents. By their very nature, indices are simple, and combine many details into a generalized, overall description of the atmosphere or ocean which can be used to characterize the factors which impact the global climate system.

El Niño – Southern Oscillation

El Niño-Southern Oscillation (ENSO) is a global coupled ocean-atmosphere phenomenon. The Pacific ocean signatures, El Niño and La Niña are important temperature fluctuations in surface waters of the tropical Eastern Pacific Ocean. The name El Niño, from the Spanish for "the little boy", refers to the Christ child, because the phenomenon is usually noticed around Christmas time in the Pacific Ocean off the west coast of South America.[3] La Niña means "the little girl".[4] Their effect on climate in the subtropics and the tropics are profound. The atmospheric signature, the Southern Oscillation (SO) reflects the monthly or seasonal fluctuations in the air pressure difference between Tahiti and Darwin. The most recent occurrence of El Niño started in September 2006[5] and lasted until early 2007.[6] ENSO is a set of interacting parts of a single global system of coupled ocean-atmosphere climate fluctuations that come about as a consequence of oceanic and atmospheric circulation. ENSO is the most prominent known source of inter-annual variability in weather and climate around the world. The cycle occurs every two to seven years, with El Niño lasting nine months to two years within the longer term cycle,[7] though not all areas globally are affected. ENSO has signatures in the Pacific, Atlantic and Indian Oceans.

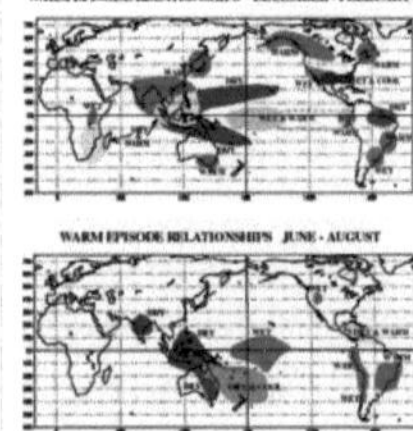

El Niño impacts

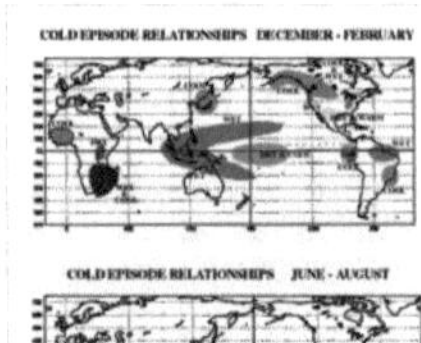

La Niña impacts

In the Pacific, during major warm events, El Niño warming extends over much of the tropical Pacific and becomes clearly linked to the SO intensity. While ENSO events are basically in phase between the Pacific and Indian Oceans, ENSO events in the Atlantic Ocean lag behind those in the Pacific by 12–18 months. Many of the countries most affected by ENSO events are developing countries within tropical sections of continents with economies that are largely dependent upon their agricultural and fishery sectors as a major source of food supply, employment, and foreign exchange.[8] New capabilities to predict the onset of ENSO events in the three oceans can have global socio-economic impacts. While ENSO is a global and natural part of the Earth's climate, whether its intensity or frequency may change as a result of global warming is an important concern. Low-frequency variability has been evidenced: the quasi-decadal oscillation (QDO). Inter-decadal (ID) modulation of ENSO (from PDO or IPO) might exist. This could explain the so-called protracted ENSO of the early 1990s.

Madden–Julian Oscillation

The Madden–Julian Oscillation (MJO) is an equatorial traveling pattern of anomalous rainfall that is planetary in scale. It is characterized by an eastward progression of large regions of both enhanced and suppressed tropical rainfall, observed mainly over the Indian and Pacific Oceans. The anomalous rainfall is usually first evident over the western Indian Ocean, and remains evident as it propagates over the very warm ocean waters of the western and central tropical Pacific. This pattern of tropical rainfall then generally becomes very nondescript as it moves over the cooler ocean waters of the eastern Pacific but reappears over the tropical Atlantic and Indian Oceans. The wet phase of enhanced convection and precipitation is followed by a dry phase where convection is suppressed. Each cycle lasts approximately 30–60 days. The MJO is also known as the 30–60 day oscillation, 30–60 day wave, or intraseasonal oscillation.

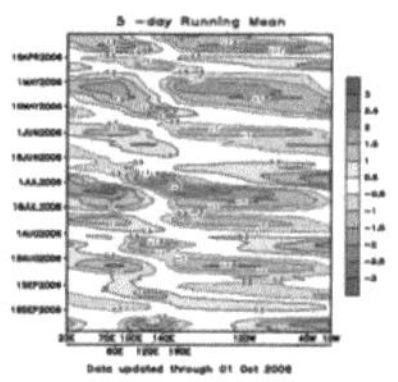

Note how the MJO moves eastward with time.

North Atlantic Oscillation (NAO)

Indices of the NAO are based on the difference of normalized sea level pressure (SLP) between Ponta Delgada, Azores and Stykkisholmur/Reykjavik, Iceland. The SLP anomalies at each station were normalized by division of each seasonal mean pressure by the long-term mean (1865–1984) standard deviation. Normalization is done to avoid the series of being dominated by the greater variability of the northern of the two stations. Positive values of the index indicate stronger-than-average westerlies over the middle latitudes.[9]

Northern Annular Mode (NAM) or Arctic Oscillation (AO)

The NAM, or AO, is defined as the first EOF of northern hemisphere winter SLP data from the tropics and subtropics. It explains 23% of the average winter (December–March) variance, and it is dominated by the NAO structure in the Atlantic. Although there are some subtle differences from the regional pattern over the Atlantic and Arctic, the main difference is larger amplitude anomalies over the North Pacific of the same sign as those over the Atlantic. This feature gives the NAM a more annular (or zonally symmetric) structure.[9]

Northern Pacific (NP) Index

The NP Index is the area-weighted sea level pressure over the region 30N–65N, 160E–140W.[9]

Pacific Decadal Oscillation (PDO)

The PDO is a pattern of Pacific climate variability that shifts phases on at least inter-decadal time scale, usually about 20 to 30 years. The PDO is detected as warm or cool surface waters in the Pacific Ocean, north of 20° N. During a "warm", or "positive", phase, the west Pacific becomes cool and part of the eastern ocean warms; during a "cool" or "negative" phase, the opposite pattern occurs. The mechanism by which the pattern lasts over several years has not been identified; one suggestion is that a thin layer of warm water during summer may shield deeper cold waters. A PDO signal has been reconstructed to 1661 through tree-ring chronologies in the Baja California area.

Interdecadal Pacific Oscillation (IPO)

The Interdecadal Pacific Oscillation (IPO or ID) display similar sea surface temperature (SST) and sea level pressure patterns to the PDO, with a cycle of 15–30 years, but affects both the north and south Pacific. In the tropical Pacific, maximum SST anomalies are found away from the equator. This is quite different from the quasi-decadal oscillation (QDO) with a period of 8–12 years and maximum SST anomalies straddling the equator, thus resembling ENSO.

Models

Climate models use quantitative methods to simulate the interactions of the atmosphere, oceans, land surface, and ice. They are used for a variety of purposes from study of the dynamics of the weather and climate system to projections of future climate. All climate models balance, or very nearly balance, incoming energy as short wave (including visible) electromagnetic radiation to the earth with outgoing energy as long wave (infrared) electromagnetic radiation from the earth. Any unbalance results in a change in the average temperature of the earth.

The most talked-about models of recent years have been those relating temperature to emissions of carbon dioxide (see greenhouse gas). These models predict an upward trend in the surface temperature record, as well as a more rapid increase in temperature at higher latitudes.

Models can range from relatively simple to quite complex:

- A simple radiant heat transfer model that treats the earth as a single point and averages outgoing energy
- this can be expanded vertically (radiative-convective models), or horizontally
- finally, (coupled) atmosphere–ocean–sea ice **global climate models** discretise and solve the full equations for mass and energy transfer and radiant exchange.

Differences with meteorology

In contrast to meteorology, which focuses on short term weather systems lasting up to a few weeks, climatology studies the frequency and trends of those systems. It studies the periodicity of weather events over years to millennia, as well as changes in long-term average weather patterns, in relation to atmospheric conditions. Climatologists, those who practice climatology, study both the nature of climates – local, regional or global – and the natural or human-induced factors that cause climates to change. Climatology considers the past and can help predict future climate change.

Phenomena of climatological interest include the atmospheric boundary layer, circulation patterns, heat transfer (radiative, convective and latent), interactions between the atmosphere and the oceans and land surface (particularly vegetation, land use and topography), and the chemical and physical composition of the atmosphere.

Use in weather forecasting

A more complicated way of making a forecast, the analog technique requires remembering a previous weather event which is expected to be mimicked by an upcoming event. What makes it a difficult technique to use is that there is rarely a perfect analog for an event in the future.[10] Some call this type of forecasting **pattern recognition**, which remains a useful method of observing rainfall over data voids such as oceans with knowledge of how satellite imagery relates to precipitation rates over land,[11] as well as the forecasting of precipitation amounts and distribution in the future. A variation on this theme is used in Medium Range forecasting, which is known as teleconnections, when you use systems in other locations to help pin down the location of another system within the surrounding regime.[12] One method of using teleconnections are by using climate indices such as ENSO-related phenomena.[13]

See also

- Biogeochemistry
- Climate
- Climate Prediction Center
- Edmund Halley
- Geophysics
- Helmut Landsberg
- List of climate scientists
- Meteorology
- National Climatic Data Center
- Paleoclimatology
- Paleotempestology
- Pangaea Expedition
- Tornado climatology
- Tropical cyclone rainfall climatology
- Urban climatology

References

[1] Climate Prediction Center. Climate Glossary. (http://www.cpc.noaa.gov/products/outreach/glossary.shtml#C) Retrieved on November 23, 2006.

[2] Life Stories. Francis Galton. (http://www.channel4.com/science/microsites/S/science/life/biog_galton.html) Retrieved on April 19, 2007.

[3] California Department of Fish and Game, Marine Region. El Niño Information. (http://www.dfg.ca.gov/mrd/elnino.html) Retrieved on June 7, 2007.

[4] La Niña. (http://library.advanced.org/20901/la_nina.htm)

[5] El Nino forms in Pacific Ocean (http://edition.cnn.com/2006/WEATHER/09/13/weather.nino.reut/index.html), CNN

[6] "There Goes El Nino, Here Comes La Nina" (http://www.cbsnews.com/stories/2007/02/28/tech/main2523483.shtml). The Associated Press / CBS News. February 28, 2007. . Retrieved March 2, 2007.

[7] Climate Prediction Center (December 19, 2005). "ENSO FAQ: How often do El Niño and La Niña typically occur?" (http://www.cpc.noaa.gov/products/analysis_monitoring/ensostuff/ensofaq.shtml#HOWOFTEN). National Centers for Environmental Prediction. . Retrieved July 26, 2009.

[8] Pearcy, W. G.; Schoener, A. (1987). "Changes in the marine biota coincident with the 1982–83 El Niño in the northeastern subarctic Pacific Ocean" (http://www.agu.org/pubs/crossref/1987/JC092iC13p14417.shtml). *Journal of Geophysical Research* **92** (C13): 14417–14428. Bibcode 1987JGR....9214417P. doi:10.1029/JC092iC13p14417. .

[9] National Center for Atmospheric Research. Climate Analysis Section. (http://www.cgd.ucar.edu/cas/jhurrell/indices.info.html) Retrieved on June 7, 2007.

[10] Other Forecasting Methods: climatology, analogue and numerical weather prediction. (http://ww2010.atmos.uiuc.edu/(Gh)/guides/mtr/fcst/mth/oth.rxml) Retrieved on February 16, 2006.

[11] Kenneth C. Allen. Pattern Recognition Techniques Applied to the NASA-ACTS Order-Wire Problem. (http://stinet.dtic.mil/oai/oai?&verb=getRecord&metadataPrefix=html&identifier=ADP006190) Retrieved on February 16, 2007.

[12] Weather Associates, Inc. The Role of Teleconnections & Ensemble Forecasting in Extended- to Medium-Range Forecasting. (http://www.weatherassociates.com/courses.htm) Retrieved on February 16, 2007.

[13] Thinkquest.org. Teleconnections: Linking El Niño with Other Places. (http://library.thinkquest.org/20901/teleconnections.htm) Retrieved on February 16, 2007.

External links

- The Earth's climate – Centre national de la recherche scientifique (CNRS – France) (http://www.cnrs.fr/climate)
- Climatology News (http://www.climatologynews.com) Daily publication with news in all areas of climatology plus free news feeds for webmasters.
- Climate Prediction Center (http://www.cpc.ncep.noaa.gov)
- KNMI Climate Explorer (http://climexp.knmi.nl) The Royal Netherlands Meteorological Institute's Climate Explorer graphs climatological relationships of spatial and temporal data.
- Climatology as a Profession (http://www.aip.org/history/climate/climogy.htm) Amer. Inst. of Physics account of the history of the discipline of climatology in the 20th century

Meteorology

Meteorology is the interdisciplinary scientific study of the atmosphere. Studies in the field stretch back millennia, though significant progress in meteorology did not occur until the eighteenth century. The nineteenth century saw breakthroughs occur after observing networks developed across several countries. After the development of the computer in the latter half of the twentieth century breakthroughs in weather forecasting were achieved.

Meteorological phenomena are observable weather events which illuminate and are explained by the science of meteorology. Those events are bound by the variables that exist in Earth's atmosphere; temperature, air pressure, water vapor, and the gradients and interactions of each variable, and how they change in time. Different spatial scales are studied to determine how systems on local, region, and global levels impact weather and climatology.

Meteorology, climatology, atmospheric physics, and atmospheric chemistry are sub-disciplines of the atmospheric sciences. Meteorology and hydrology compose the interdisciplinary field of hydrometeorology. Interactions between Earth's atmosphere and the oceans are part of coupled ocean-atmosphere studies. Meteorology has application in many diverse fields such as the military, energy production, transport, agriculture and construction.

The word "meteorology" is from Greek μετέωρος *metéōros* "lofty; high (in the sky)" (from μετα- *meta-* "above" and ἐωρ *eōr* "to lift up") and -λογία *-logia* "-(o)logy".

History

Parhelion (sundog) at Savoie

The beginnings of meteorology can be traced back in ancient India to 3000 B.C.E [1] , such as the Upanishads, contain serious discussion about the processes of cloud formation and rain and the seasonal cycles caused by the movement of earth round the sun. Varahamithra's classical work, Brihatsamhita , written about 500 A.D. [1] , provides a clear evidence that a deep knowledge of atmospheric processes existed even in those times. In 350 BC, Aristotle wrote *Meteorology*.[2] Aristotle is considered the founder of meteorology.[3] One of the most impressive achievements described in the *Meteorology* is the description of what is now known as the hydrologic cycle.[4] The Greek scientist Theophrastus compiled a book on weather forecasting, called the *Book of Signs*. The work of Theophrastus remained a dominant influence in the study of weather and in weather forecasting for nearly 2,000 years.[5] In 25 AD, Pomponius Mela, a geographer for the Roman Empire, formalized the climatic zone system.[6] Around the 9th century, Al-Dinawari, a Kurdish naturalist, writes the *Kitab al-Nabat* (*Book of Plants*), in which he deals with the application of meteorology to agriculture during the Muslim Agricultural Revolution. He describes the meteorological character of the sky, the planets and constellations, the sun and moon, the lunar phases indicating seasons and rain, the *anwa* (heavenly bodies of rain), and atmospheric phenomena such as winds, thunder, lightning, snow, floods, valleys, rivers, lakes, wells and other sources of water.[7]

Research of visual atmospheric phenomena

In 1021, Ibn al-Haytham (Alhazen) wrote on the atmospheric refraction of light.[8] He showed that the twilight is due to atmospheric refraction and only begins when the Sun is 19 degrees below the horizon, and uses a complex geometric demonstration to measure the height of the Earth's atmosphere as 52,000 *passuum* (49 miles (79 km)),[9] [10] which is very close to the modern measurement of 50 miles (80 km). He also realized that the atmosphere also reflects light, from his observations of the sky brightening even before the Sun rises.[11]

Twilight at Baker Beach

St. Albert the Great was the first to propose that each drop of falling rain had the form of a small sphere, and that this form meant that the rainbow was produced by light interacting with each raindrop.[12] Roger Bacon was the first to calculate the angular size of the rainbow. He stated that the rainbow summit can not appear higher than 42 degrees above the horizon.[13] In the late 13th century and early 14th century, Theodoric of Freiberg and Kamāl al-Dīn al-Fārisī continued the work of Ibn al-Haytham, and they were the first to give the correct explanations for the primary rainbow phenomenon. Theoderic went further and also explained the secondary rainbow [14] In 1716, Edmund Halley suggests that aurorae are caused by "magnetic effluvia" moving along the Earth's magnetic field lines.

Instruments and classification scales

In 1441, King Sejongs son, Prince Munjong, invented the first standardized rain gauge. These were sent throughout the Joseon Dynasty of Korea as an official tool to assess land taxes based upon a farmer's potential harvest. In 1450, Leone Battista Alberti developed a swinging-plate anemometer, and is known as the first *anemometer*.[15] In 1607, Galileo Galilei constructs a thermoscope. In 1611, Johannes Kepler writes the first scientific treatise on snow crystals: "Strena Seu de Nive Sexangula (A New Year's Gift of Hexagonal Snow)".[16] In 1643, Evangelista Torricelli invents the mercury barometer.[15] In 1662, Sir Christopher Wren invented the mechanical, self-emptying, tipping bucket rain gauge. In 1714, Gabriel Fahrenheit creates a reliable scale for measuring temperature with a mercury-type thermometer.[17] In 1742, Anders Celsius, a Swedish astronomer, proposed the 'centigrade' temperature scale, the predecessor of the current Celsius scale.[18] In 1783, the first hair hygrometer is demonstrated by Horace-Bénédict de Saussure. In 1802-1803, Luke Howard writes *On the Modification of Clouds* in which he assigns cloud types Latin names.[19] In

A hemispherical cup anemometer

1806, Francis Beaufort introduced his system for classifying wind speeds.[20] Near the end of the 19th century the first cloud atlases were published, including the *International Cloud Atlas*, which has remained in print ever since. The April 1960 launch of the first successful weather satellite, TIROS-1, marked the beginning of the age where weather information became available globally.

Atmospheric composition research

In 1648, Blaise Pascal rediscovers that atmospheric pressure decreases with height, and deduces that there is a vacuum above the atmosphere.[21] In 1738, Daniel Bernoulli publishes *Hydrodynamics*, initiating the kinetic theory of gases and established the basic laws for the theory of gases.[22] In 1761, Joseph Black discovers that ice absorbs heat without changing its temperature when melting. In 1772, Black's student Daniel Rutherford discovers nitrogen, which he calls *phlogisticated air*, and together they developed the phlogiston theory.[23] In 1777, Antoine Lavoisier discovers oxygen and develops an explanation for combustion.[24] In 1783, in Lavoisier's book *Reflexions sur le phlogistique*,[25] he deprecates the phlogiston theory and proposes a caloric theory.[26] [27] In 1804, Sir John Leslie observes that a matte black surface radiates heat more effectively than a polished surface, suggesting the importance of black body radiation. In 1808, John Dalton defends caloric theory in *A New System of Chemistry* and describes how it combines with matter, especially gases; he proposes that the heat capacity of gases varies inversely with atomic weight. In 1824, Sadi Carnot analyzes the efficiency of steam engines using caloric theory; he develops the notion of a reversible process and, in postulating that no such thing exists in nature, lays the foundation for the second law of thermodynamics.

Research into cyclones and air flow

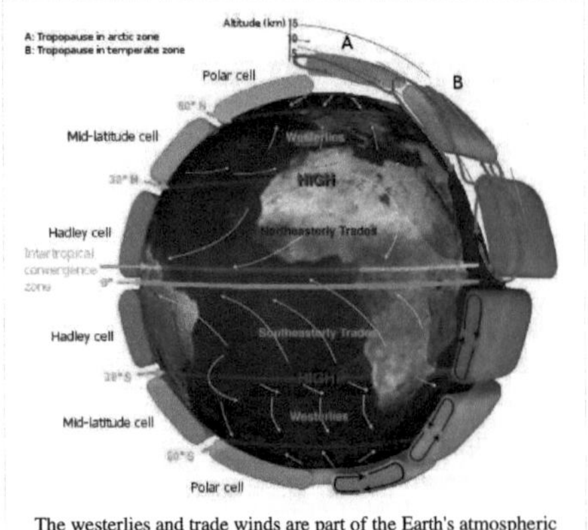

The westerlies and trade winds are part of the Earth's atmospheric circulation

In 1494, Christopher Columbus experiences a tropical cyclone, leads to the first written European account of a hurricane.[28] In 1686, Edmund Halley presents a systematic study of the trade winds and monsoons and identifies solar heating as the cause of atmospheric motions.[29] In 1735, an *ideal* explanation of global circulation through study of the Trade winds was written by George Hadley.[30] In 1743, when Benjamin Franklin is prevented from seeing a lunar eclipse by a hurricane, he decides that cyclones move in a contrary manner to the winds at their periphery.[31] Understanding the kinematics of how exactly the rotation of the Earth affects airflow was partial at first. Gaspard-Gustave Coriolis published a paper in 1835 on the energy yield of machines with rotating parts, such as waterwheels.[32] In 1856, William Ferrel proposed the existence of a circulation cell in the mid-latitudes with air being deflected by the Coriolis force to create the prevailing westerly winds.[33] Late in the 19th century the full extent of the large scale interaction of pressure gradient force and deflecting force that in the end causes air masses to move *along* isobars was understood. By 1912, this deflecting force was named the Coriolis effect.[34] Just after World War I, a group of meteorologists in Norway led by Vilhelm Bjerknes developed the Norwegian cyclone model that explains the generation, intensification and ultimate decay (the life cycle) of mid-latitude cyclones, introducing the idea of fronts, that is, sharply defined boundaries between air masses.[35] The group included Carl-Gustaf Rossby (who was the first to explain the large scale atmospheric flow in terms of fluid dynamics), Tor Bergeron (who first determined the mechanism by which rain forms) and Jacob Bjerknes.

Observation networks and weather forecasting

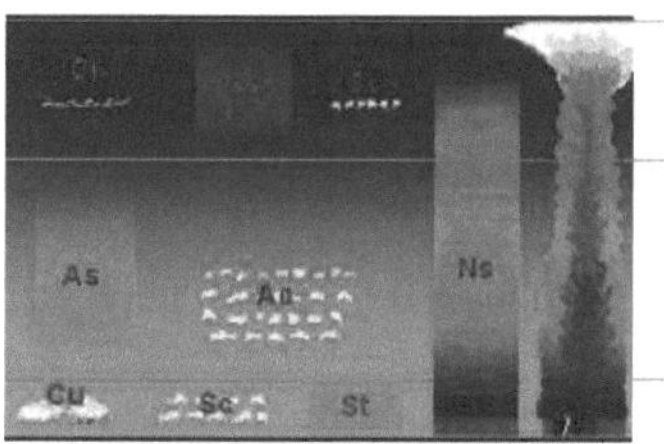

Cloud classification by altitude of occurrence

In 1654, Ferdinando II de Medici establishes the first *weather observing* network, that consisted of meteorological stations in Florence, Cutigliano, Vallombrosa, Bologna, Parma, Milan, Innsbruck, Osnabrück, Paris and Warsaw. Collected data was centrally sent to Florence at regular time intervals.[36] In 1832, an electromagnetic telegraph was created by Baron Schilling.[37] The arrival of the electrical telegraph in 1837 afforded, for the first time, a practical method for quickly gathering surface weather observations from a wide area.[38] This data could be used to produce maps of the state of the atmosphere for a region near the Earth's surface and to study how these states evolved through time. To make frequent weather forecasts based on these data required a reliable network of observations, but it was not until 1849 that the Smithsonian Institution began to establish an observation network across the United States under the leadership of Joseph Henry.[39] Similar observation networks were established in Europe at this time. In 1854, the United Kingdom government appointed Robert FitzRoy to the new office of *Meteorological Statist to the Board of Trade* with the role of gathering weather observations at sea. FitzRoy's office became the United Kingdom Meteorological Office in 1854, the first national meteorological service in the world. The first daily weather forecasts made by FitzRoy's Office were published in *The Times* newspaper in 1860. The following year a system was introduced of hoisting storm warning cones at principal ports when a gale was expected.

Over the next 50 years many countries established national meteorological services. The India Meteorological Department (1875) was established following tropical cyclone and monsoon related famines in the previous decades.[40] The Finnish Meteorological Central Office (1881) was formed from part of Magnetic Observatory of Helsinki University.[41] Japan's Tokyo Meteorological Observatory, the forerunner of the Japan Meteorological Agency, began constructing surface weather maps in 1883.[42] The United States Weather Bureau (1890) was established under the United States Department of Agriculture. The Australian Bureau of Meteorology (1906) was established by a Meteorology Act to unify existing state meteorological services.[43] [44]

Numerical weather prediction

A meteorologist at the console of the IBM 7090 in the Joint Numerical Weather Prediction Unit. c. 1965

In 1904, Norwegian scientist Vilhelm Bjerknes first argued in his paper *Weather Forecasting as a Problem in Mechanics and Physics* that it should be possible to forecast weather from calculations based upon natural laws.[45]

It was not until later in the 20th century that advances in the understanding of atmospheric physics led to the foundation of modern numerical weather prediction. In 1922, Lewis Fry Richardson published "Weather Prediction By Numerical Process", after finding notes and derivations he worked on as an ambulance driver in World War I. He described therein how small terms in the prognostic fluid dynamics equations governing atmospheric flow could be neglected, and a finite differencing scheme in time and space could be devised, to allow numerical prediction solutions to be found. Richardson envisioned a large auditorium of thousands of people performing the calculations and passing them to others. However, the sheer number of calculations required was too large to be completed without the use of computers, and the size of the grid and time steps led to unrealistic results in deepening systems. It was later found, through numerical analysis, that this was due to numerical instability.

Starting in the 1950s, numerical forecasts with computers became feasible.[46] The first weather forecasts derived this way used barotropic (that means, single-vertical-level) models, and could successfully predict the large-scale movement of midlatitude Rossby waves, that is, the pattern of atmospheric lows and highs..[47]

In the 1960s, the chaotic nature of the atmosphere was first observed and mathematically described by Edward Lorenz, founding the field of chaos theory.[48] These advances have led to the current use of ensemble forecasting in most major forecasting centers, to take into account uncertainty arising from the chaotic nature of the atmosphere.[49] Climate models have been developed that feature a resolution comparable to older weather prediction models. These climate models are used to investigate long-term climate shifts, such as what effects might be caused by human emission of greenhouse gases.

Meteorologists

Further information: Weather forecasting

Meteorologists are scientists who study meteorology.[50] Meteorologists work in government agencies, private consulting and research services, industrial enterprises, utilities, radio and television stations, and in education. In the United States, meteorologists held about 9,400 jobs in 2009.[51]

Meteorologists are best-known for forecasting the weather. Many radio and television weather forecasters are professional meteorologists, while others are merely reporters with no formal meteorological training. The American Meteorological Society and National Weather Association issue "Seals of Approval" to weather broadcasters who meet certain requirements.

Equipment

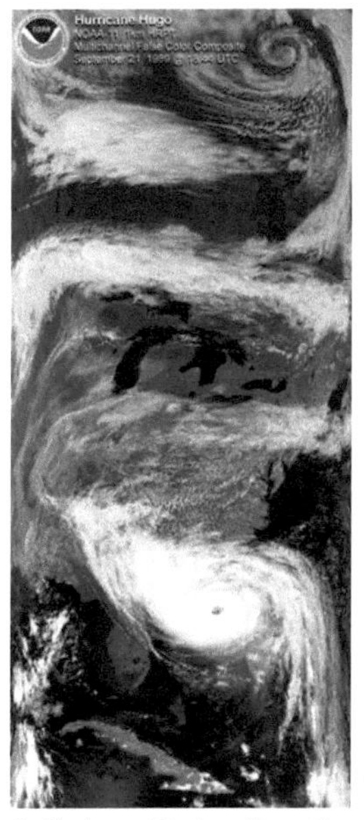

Satellite image of Hurricane Hugo with a polar low visible at the top of the image.

Each science has its own unique sets of laboratory equipment. In the atmosphere, there are many things or qualities of the atmosphere that can be measured. Rain, which can be observed, or seen anywhere and anytime was one of the first ones to be measured historically. Also, two other accurately measured *qualities* are wind and humidity. Neither of these can be *seen* but can be felt. The devices to measure these three sprang up in the mid-15th century and were respectively the rain gauge, the anemometer, and the hygrometer.[52]

Sets of surface measurements are important data to meteorologists. They give a snapshot of a variety of weather conditions at one single location and are usually at a weather station, a ship or a weather buoy. The measurements taken at a weather station can include any number of atmospheric observables. Usually, temperature, pressure, wind measurements, and humidity are the variables that are measured by a thermometer, barometer, anemometer, and hygrometer, respectively.[53] Upper air data are of crucial importance for weather forecasting. The most widely used technique is launches of radiosondes. Supplementing the radiosondes a network of aircraft collection is organized by the World Meteorological Organization.

Remote sensing, as used in meteorology, is the concept of collecting data from remote weather events and subsequently producing weather information. The common types of remote sensing are Radar, Lidar, and satellites (or photogrammetry). Each collects data about the atmosphere from a remote location and, usually, stores the data where the instrument is located. RADAR and LIDAR are not passive because both use EM radiation to illuminate a specific portion of the atmosphere.[54] Weather satellites along with more general-purpose Earth-observing satellites circling the earth at various altitudes have become an indispensable tool for studying a wide range of phenomena from forest fires to El Niño.

Spatial scales

In the study of the atmosphere, meteorology can be divided into distinct areas of emphasis depending on the temporal scope and spatial scope of interest. At one extreme of this scale is climatology. In the timescales of hours to days, meteorology separates into micro-, meso-, and synoptic scale meteorology. Respectively, the geospatial size of each of these three scales relates directly with the appropriate timescale.

Other subclassifications are available based on the need by or by the unique, local or broad effects that are studied within that sub-class.

Microscale

Microscale meteorology is the study of atmospheric phenomena of about 1 km or less. Individual thunderstorms, clouds, and local turbulence caused by buildings and other obstacles, such as individual hills fall within this category.[55]

Mesoscale

Mesoscale meteorology is the study of atmospheric phenomena that has horizontal scales ranging from microscale limits to synoptic scale limits and a vertical scale that starts at the Earth's surface and includes the atmospheric boundary layer, troposphere, tropopause, and the lower section of the stratosphere. Mesoscale timescales last from less than a day to the lifetime of the event, which in some cases can be weeks. The events typically of interest are thunderstorms, squall lines, fronts, precipitation bands in tropical and extratropical cyclones, and topographically generated weather systems such as mountain waves and sea and land breezes.[56]

Synoptic scale

Synoptic scale meteorology is generally large area dynamics referred to in horizontal coordinates and with respect to time. The phenomena typically described by synoptic meteorology include events like extratropical cyclones, baroclinic troughs and ridges, frontal zones, and to some extent jet streams. All of these are typically given on weather maps for a specific time. The minimum horizontal scale of synoptic phenomena are limited to the spacing between surface observation stations.[57]

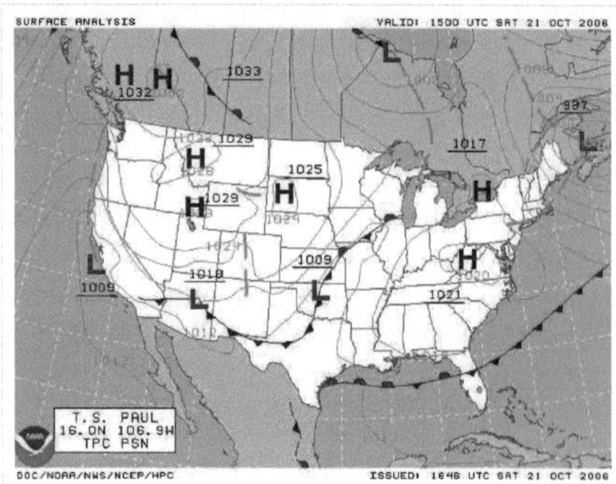

NOAA: Synoptic scale weather analysis.

Global scale

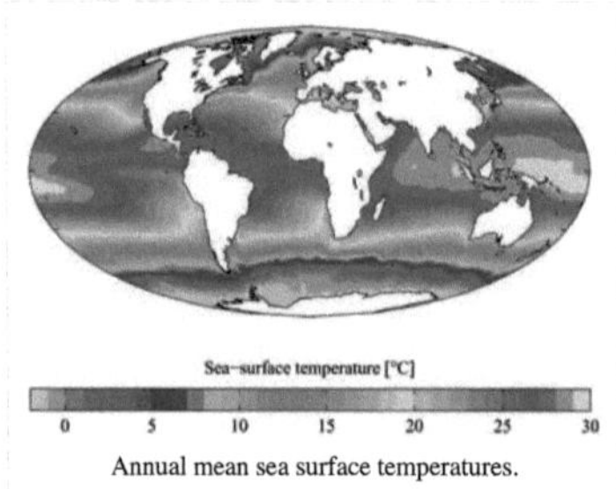

Annual mean sea surface temperatures.

Global scale meteorology is study of weather patterns related to the transport of heat from the tropics to the poles. Also, very large scale oscillations are of importance. These oscillations have time periods typically on the order of months, such as the Madden-Julian Oscillation, or years, such as the El Niño-Southern Oscillation and the Pacific decadal oscillation. Global scale pushes the thresholds of the perception of meteorology into climatology. The traditional definition of climate is pushed in to larger timescales with the further understanding of how the global oscillations cause both climate and weather disturbances in the synoptic and mesoscale timescales.

Numerical Weather Prediction is a main focus in understanding air-sea interaction, tropical meteorology, atmospheric predictability, and tropospheric/stratospheric processes.[58] The Naval Research Laboratory in Monterey produces the atmospheric model called **NOGAPS**, a global scale atmospheric model, this model is run operationally at Fleet Numerical Meteorology and Oceanography Center. Many other global atmospheric models are run by national meteorological agencies.

Some meteorological principles

Boundary layer meteorology

Boundary layer meteorology is the study of processes in the air layer directly above Earth's surface, known as the atmospheric boundary layer (ABL). The effects of the surface – heating, cooling, and friction – cause turbulent mixing within the air layer. Significant fluxes of heat, matter, or momentum on time scales of less than a day are advected by turbulent motions.[59] Boundary layer meteorology includes the study of all types of surface-atmosphere boundary, including ocean, lake, urban land and non-urban land.

Dynamic meteorology

Dynamic meteorology generally focuses on the fluid dynamics of the atmosphere. The idea of air parcel is used to define the smallest element of the atmosphere, while ignoring the discrete molecular and chemical nature of the atmosphere. An air parcel is defined as a point in the fluid continuum of the atmosphere. The fundamental laws of fluid dynamics, thermodynamics, and motion are used to study the atmosphere. The physical quantities that characterize the state of the atmosphere are temperature, density, pressure, etc. These variables have unique values in the continuum.[60]

Applications

Weather forecasting

Weather forecasting is the application of science and technology to predict the state of the atmosphere for a future time and a given location. Human beings have attempted to predict the weather informally for millennia, and formally since at least the nineteenth century.[61] [62] Weather forecasts are made by collecting quantitative data about the current state of the atmosphere and using scientific understanding of atmospheric processes to project how the atmosphere will evolve.[63]

Forecast of surface pressures five days into the future for the north Pacific, North America, and north Atlantic Ocean

Once an all-human endeavor based mainly upon changes in barometric pressure, current weather conditions, and sky condition,[64] [65] forecast models are now used to determine future conditions. Human input is still required to pick the best possible forecast model to base the forecast upon, which involves pattern recognition skills, teleconnections, knowledge of model performance, and knowledge of model biases. The chaotic nature of the atmosphere, the massive computational power required to solve the equations that describe the atmosphere, error involved in measuring the initial conditions, and an incomplete understanding of atmospheric processes mean that forecasts become less accurate as the difference in current time and the time for which the forecast is being made (the *range* of the forecast) increases. The use of ensembles and model consensus help narrow the error and pick the most likely outcome.[66] [67] [68]

There are a variety of end uses to weather forecasts. Weather warnings are important forecasts because they are used to protect life and property.[69] Forecasts based on temperature and precipitation are important to agriculture,[70] [71] [72] [73] and therefore to commodity traders within stock markets. Temperature forecasts are used by utility companies to estimate demand over coming days.[74] [75] [76] On an everyday basis, people use weather forecasts to determine what to wear on a given day. Since outdoor activities are severely curtailed by heavy rain, snow and the wind chill, forecasts can be used to plan activities around these events, and to plan ahead and survive them.

Aviation meteorology

Aviation meteorology deals with the impact of weather on air traffic management. It is important for air crews to understand the implications of weather on their flight plan as well as their aircraft, as noted by the Aeronautical Information Manual[77] :

> *The effects of ice on aircraft are cumulative-thrust is reduced, drag increases, lift lessens, and weight increases. The results are an increase in stall speed and a deterioration of aircraft performance. In extreme cases, 2 to 3 inches of ice can form on the leading edge of the airfoil in less than 5 minutes. It takes but 1/2 inch of ice to reduce the lifting power of some aircraft by 50 percent and increases the frictional drag by an equal percentage.*[78]

Agricultural meteorology

Meteorologists, soil scientists, agricultural hydrologists, and agronomists are persons concerned with studying the effects of weather and climate on plant distribution, crop yield, water-use efficiency, phenology of plant and animal development, and the energy balance of managed and natural ecosystems. Conversely, they are interested in the role of vegetation on climate and weather.[79]

Hydrometeorology

Hydrometeorology is the branch of meteorology that deals with the hydrologic cycle, the water budget, and the rainfall statistics of storms.[80] A hydrometeorologist prepares and issues forecasts of accumulating (quantitative) precipitation, heavy rain, heavy snow, and highlights areas with the potential for flash flooding. Typically the range of knowledge that is required overlaps with climatology, mesoscale and synoptic meteorology, and other geosciences.[81]

Nuclear meteorology

Nuclear meteorology investigates the distribution of radioactive aerosols and gases in the atmosphere.[82]

Maritime meteorology

Maritime meteorology deals with air and wave forecasts for ships operating at sea. Organizations such as the Ocean Prediction Center, Honolulu National Weather Service forecast office, United Kingdom Met Office, and JMA prepare high seas forecasts for the world's oceans.

See also

- Aerography
- American Practical Navigator
- Atmospheric circulation
- Atmospheric layers
- Atmospheric models
- Atmospheric thermodynamics
- Eddy covariance flux (aka, eddy correlation, eddy flux)
- ENSO (El Niño-Southern Oscillation)
- Index of meteorology articles
- List of weather instruments
- List of meteorology institutions
- List of Russian meteorologists
- Madden-Julian oscillation
- Meteorological Winter
- National Weatherperson's Day
- Space weather
- Walker circulation

References

[1] http://www.imd.gov.in/doc/history/history.htm

[2] Development of Meteorology (http://www.infoplease.com/ce6/weather/A0859595.html)

[3] Meteorology by Lisa Alter (http://yale.edu/ynhti/curriculum/units/1994/5/94.05.01.x.html)

[4] Aristotle (2004) [350 B.C.E]. *Meteorology* (http://etext.library.adelaide.edu.au/a/aristotle/meteorology/). The University of Adelaide Library, University of Adelaide, South Australia 5005: eBooks@Adelaide. . "Translated by E. W. Webster"

[5] Weather: Forecasting from the Beginning (http://www.infoplease.com/cig/weather/forecasting-from-beginning.html)

[6] "Timeline of geography, paleontology" (http://www.paleorama.com/timelines/geography.html). Paleorama.com. . "Following the path of Discovery"

[7] Fahd, Toufic. "Botany and agriculture". pp. 815, in Morelon, Régis; Rashed, Roshdi (1996). *Encyclopedia of the History of Arabic Science*. **3**. Routledge. ISBN 0415124107

[8] Dr. Mahmoud Al Deek. "Ibn Al-Haitham: Master of Optics, Mathematics, Physics and Medicine, *Al Shindagah*, November–December 2004.

[9] Frisinger, H. Howard (March 1973). "Aristotle's Legacy in Meteorology". *Bulletin of the American Meteorological Society* **3** (3): 198–204 [201]

[10] George Sarton, *Introduction to the History of Science* (cf. Dr. A. Zahoor and Dr. Z. Haq (1997), Quotations from Famous Historians of Science (http://www.cyberistan.org/islamic/Introl1.html))

[11] Bradley Steffens (2006), *Ibn al-Haytham: First Scientist*, Chapter Five (http://www.ibnalhaytham.net/custom.em?pid=673906), Morgan Reynolds Publishing, ISBN 1-59935-024-6

[12] Ancient and pre-Renaissance contributors to Meteorology (http://rammb.cira.colostate.edu/dev/hillger/ancient.htm#magnus), National Oceanic and Atmospheric Administration

[13] The rainbow bridge: rainbows in art, myth, and science (http://books.google.es/books?id=kZcCtT1ZeaEC&pg=PA155&dq=Rainbow+roger+Bacon&hl=es&ei=Ion5S6DMDMP88Aaxv8SrBw&sa=X&oi=book_result&ct=result&resnum=8&ved=0CE0Q6AEwBw#v=onepage&q=Rainbow roger Bacon&f=false), p.156,Raymond L. Lee,Alistair B. Fraser

[14] Topdemir, Hüseyin Gazi (2007), Kamal Al-din Al-Farisi´s explanation of the rainbow (http://www.idosi.org/hssj/hssj2(1)07/10.pdf)

[15] Jacobson, Mark Z. (June 2005) (paperback). *Fundamentals of Atmospheric Modeling* (2nd ed.). New York: Cambridge University Press. pp. 828. ISBN 9780521548656.

[16] Highlights in the study of snowflakes and snow crystals (http://www.its.caltech.edu/~atomic/snowcrystals/earlyobs/earlyobs.htm)

[17] Grigull, U., Fahrenheit, a Pioneer of Exact Thermometry. Heat Transfer, 1966, The Proceedings of the 8th International Heat Transfer Conference, San Francisco, 1966, Vol. 1.

[18] Beckman, Olof, History of the Celsius temperature scale. (http://www.astro.uu.se/history/Celsius_scale.html), *translated*, Anders Celsius (Elementa,84:4,2001); *English*

[19] Thornes, John. E. (1999). *John Constable's Skies*. The University of Birmingham Press, pp. 189. ISBN 1-902459-02-4.

[20] Bill Giles O.B.E. (2009). Beaufort Scale. (http://www.bbc.co.uk/weather/features/understanding/beaufort_scale.shtml) BBC. Retrieved on 2009-05-12.

[21] Florin to Pascal, September 1647,*Œuves completes de Pascal*, 2:682.

[22] O'Connor, John J.; Robertson, Edmund F., "Meteorology" (http://www-history.mcs.st-andrews.ac.uk/Biographies/Bernoulli_Daniel.html), *MacTutor History of Mathematics archive*, University of St Andrews, .

[23] Biographical note at "Lectures and Papers of Professor Daniel Rutherford (1749–1819), and Diary of Mrs Harriet Rutherford" (http://www.londonmet.ac.uk/genesis/search/$-search-results.cfm?CCODE=2476).

[24] "Sur la combustion en général" ("On Combustion in general," 1777) and "Considérations Générales sur la Nature des Acides" ("General Considerations on the Nature of Acids," 1778).

[25] Lavoisier, ("Reflections on Phlogiston," 1783).

[26] Lavoisier, Antoine, *Elements of Chemistry*, Dover Publications Inc., New York, NY,1965, 511 pages.

[27] The 1880 edition of A Guide to the Scientific Knowledge of Things Familiar, a 19th century educational science book, explained heat transfer in terms of the flow of caloric.

[28] Morison, Samuel Eliot,*Admiral of the Ocean Sea: A Life of Cristopher Columbus*, Boston, 1942, page 617.
[29] Cook, Alan H., *Edmond Halley: Charting the Heavens and the Seas* (Oxford: Clarendon Press, 1998)
[30] George Hadley, "Concerning the cause of the general trade winds," Philosophical Transactions, vol. 39 (1735).
[31] Dorst, Neal, FAQ:_Hurricanes,_Typhoons,_and_Tropical_Cyclones:_Hurricane_Timeline (http://www.aoml.noaa.gov/hrd/tcfaq/J6.html), Hurricane_Research_Division,_Atlantic_Oceanographic_and_Meteorological_Laboratory,_NOAA (http://www.aoml.noaa.gov/hrd/), *January 2006*.
[32] G-G Coriolis (1835). "Sur les équations du mouvement relatif des systèmes de corps". *J. De l'Ecole royale polytechnique* **15**: 144–154.
[33] William Ferrel. An Essay on the Winds and the Currents of the Ocean. (http://www.aos.princeton.edu/WWWPUBLIC/gkv/history/ferrel-nashville56.pdf) Retrieved on 2009-01-01.
[34] Arthur Gordon Webster (1912). *The Dynamics of Particles and of Rigid, Elastic, and Fluid Bodies* (http://books.google.com/?id=zXkRAAAAYAAJ&pg=PA320&dq=coriolis+centrifugal+force+date:0-1920). B. G. Teubner. p. 320. .
[35] Shaye Johnson. The Norwegian Cyclone Model. (http://weather.ou.edu/~metr4424/Files/Norwegian_Cyclone_Model.pdf#search="norwegian cyclone model") Retrieved on 2006-10-11.
[36] Raymond S. Bradley, Philip D. Jones, *Climate Since A.D. 1500*, Routledge, 1992, ISBN 0-415-07593-9, p.144
[37] Rebecca Martin (2009). Catapult - Indepth - Communication: telegraph. (http://www.abc.net.au/cgi-bin/common/printfriendly.pl?/catapult/indepth/telegraph.htm) Australian Broadcasting Corporation. Retrieved on 2009-05-12.
[38] Library of Congress. The Invention of the Telegraph. (http://memory.loc.gov/ammem/sfbmhtml/sfbmtelessay.html) Retrieved on 2009-01-01.
[39] Smithsonian Institution Archives (http://www.si.edu/archives/ihd/jhp/joseph03.htm)
[40] India Meteorological Department Establishment of IMD. (http://www.imd.gov.in/doc/history/eastablishment-of-imd.htm) Retrieved on 2009-01-01.
[41] Finnish Meteorological Institute. History of Finnish Meteorological Institute. (http://www.fmi.fi/organization/history.html) Retrieved on 2009-01-01.
[42] Japan Meteorological Agency. History. (http://www.jma.go.jp/jma/en/History/indexe_his.htm) Retrieved on 2006-10-22.
[43] "BOM celebrates 100 years" (http://www.abc.net.au/news/stories/2008/01/01/2129737.htm). Australian Broadcasting Corporation. . Retrieved 2008-01-01.
[44] "Collections in Perth: 20. Meteorology" (http://www.naa.gov.au/naaresources/Publications/research_guides/guides/perth/chapter20.htm). National Archives of Australia. . Retrieved 2008-05-24.
[45] "Pioneers in Modern Meteorology and Climatology: Vilhelm and Jacob Bjerknes" (http://docs.lib.noaa.gov/rescue/Bibliographies/Bjerknes/Bjerknes_July_2004.pdf) (PDF). . Retrieved 2008-10-13.
[46] American Institute of Physics. Atmospheric General Circulation Modeling. (http://www.aip.org/history/sloan/gcm/) Retrieved on 2008-01-13.
[47] Cox, John D. (2002). *Storm Watchers*. John Wiley & Sons, Inc.. p. 208. ISBN 047138108X.
[48] Edward N. Lorenz, "Deterministic non-periodic flow", *Journal of the Atmospheric Sciences*, vol. 20, pages 130–141 (1963).
[49] Manousos, Peter (2006-07-19). "Ensemble Prediction Systems" (http://www.hpc.ncep.noaa.gov/ensembletraining). Hydrometeorological Prediction Center. . Retrieved 2010-12-31.
[50] Glossary of Meteorology (2009). Meteorologist. (http://amsglossary.allenpress.com/glossary/search?p=1&query=meteorologist&submit=Search) American Meteorological Society. Retrieved on 2009-05-10.
[51] Bureau of labor statistics: "Occupational Outlook Handbook, 2010-11 Edition" (http://www.bls.gov/oco/ocos051.htm)
[52] Many attempts had been made prior to the 15th century to construct adequate equipment to measure the many atmospheric variables. Many were faulty in some way or were simply not reliable. Even Aristotle notes this in some of his work; *as the difficulty* to measure the air.
[53] Office of the Federal Coordinator of Meteorology. Federal Meteorological Handbook No. 1 - Surface Weather Observations and Reports: September 2005. (http://www.ofcm.gov/fmh-1/fmh1.htm) Retrieved on 2009-01-02.
[54] Peebles, Peyton, [1998], *Radar Principles*, John Wiley & Sons, Inc., New York, ISBN 0-471-25205-0.
[55] "AMS Glossary of Meteorology" (http://amsglossary.allenpress.com/glossary/search?query=micrometeorology). *Micrometeorology*. American Meteorological Society. . Retrieved 2008-04-12.
[56] *Online Glossary of Meteorology* (http://amsglossary.allenpress.com/glossary), American Meteorological Society (http://www.ametsoc.org/) ,2nd Ed., 2000, Allen Press (http://www.allenpress.com).
[57] Bluestein, H., *Synoptic-Dynamic Meteorology in Midlatitudes: Principles of Kinematics and Dynamics, Vol. 1*, Oxford University Press, 1992; ISBN 0-19-506267-1
[58] Global Modelling (http://www.nrlmry.navy.mil/sec7532.htm), US Naval Research Laboratory, Monterrey, Ca.
[59] *Garratt, J.R.,* The atmospheric boundary layer, *Cambridge University Press, 1992; ISBN 0-521-38052-9.*
[60] Holton, J.R. [2004]. An Introduction to Dynamic Meteorology, 4th Ed., Burlington, Md: Elsevier Inc.. ISBN 0-12-354015-1.
[61] Mistic House. Astrology Lessons, History, Prediction, Skeptics, and Astrology Compatibility. (http://www.mistichouse.com/astrology-lessons.htm) Retrieved on 2008-01-12.
[62] Eric D. Craft. An Economic History of Weather Forecasting. (http://eh.net/encyclopedia/article/craft.weather.forcasting.history) Retrieved on 2007-04-15.
[63] NASA. Weather Forecasting Through the Ages. (http://earthobservatory.nasa.gov/Library/WxForecasting/wx2.html) Retrieved on 2008-05-25.

[64] Weather Doctor. Applying The Barometer To Weather Watching. (http://www.islandnet.com/~see/weather/eyes/barometer3.htm) Retrieved on 2008-05-25.

[65] Mark Moore. Field Forecasting - A Short Summary. (http://www.nwac.us/education_resources/Field_forecasting.pdf) Retrieved on 2008-05-25.

[66] Klaus Weickmann, Jeff Whitaker, Andres Roubicek and Catherine Smith. The Use of Ensemble Forecasts to Produce Improved Medium Range (3-15 days) Weather Forecasts. (http://www.cdc.noaa.gov/spotlight/12012001/) Retrieved on 2007-02-16.

[67] Todd Kimberlain. Tropical cyclone motion and intensity talk (June 2007). (http://www.hpc.ncep.noaa.gov/research/TropicalTalk.ppt) Retrieved on 2007-07-21.

[68] Richard J. Pasch, Mike Fiorino, and Chris Landsea. TPC/NHC'S REVIEW OF THE NCEP PRODUCTION SUITE FOR 2006. (http://www.emc.ncep.noaa.gov/research/NCEP-EMCModelReview2006/TPC-NCEP2006.ppt) Retrieved on 2008-05-05.

[69] National Weather Service. National Weather Service Mission Statement. (http://www.weather.gov/mission.shtml) Retrieved on 2008-05-25.

[70] Blair Fannin. Dry weather conditions continue for Texas. (http://southwestfarmpress.com/news/061406-Texas-weather/) Retrieved on 2008-05-26.

[71] Dr. Terry Mader. Drought Corn Silage. (http://beef.unl.edu/stories/200004030.shtml) Retrieved on 2008-05-26.

[72] Kathryn C. Taylor. Peach Orchard Establishment and Young Tree Care. (http://pubs.caes.uga.edu/caespubs/pubcd/C877.htm) Retrieved on 2008-05-26.

[73] Associated Press. After Freeze, Counting Losses to Orange Crop. (http://query.nytimes.com/gst/fullpage.html?res=9D0CE5DB1E30F937A25752C0A967958260) Retrieved on 2008-05-26.

[74] The New York Times. FUTURES/OPTIONS; Cold Weather Brings Surge In Prices of Heating Fuels. (http://query.nytimes.com/gst/fullpage.html?res=9F0CE7D9123AF935A15751C0A965958260) Retrieved on 2008-05-25.

[75] BBC. Heatwave causes electricity surge. (http://news.bbc.co.uk/1/hi/uk/5212724.stm) Retrieved on 2008-05-25.

[76] Toronto Catholic Schools. The Seven Key Messages of the Energy Drill Program. (http://www.tcdsb.org/environment/energydrill/EDSP_KeyMessages_FINAL.pdf) Retrieved on 2008-05-25.

[77] An international version called the Aeronautical Information Publication contains parallel information, as well as specific information on the international airports for use by the international community.

[78] *"7-1-22. PIREPs Relating to Airframe Icing"*, [February 16, 2006], Aeronautical Information Manual, FAA AIM Online (http://www.faa.gov/air_traffic/publications/ATpubs/AIM/)

[79] Agricultural and Forest Meteorology (http://www.elsevier.com/wps/find/journaldescription.cws_home/503295/description?navopenmenu=-2), Elsevier, ISSN: 0168-1923.

[80] Encyclopedia Britannica (http://www.britannica.com/eb/article-9041744/hydrometeorology), 2007.

[81] About the HPC (http://www.hpc.ncep.noaa.gov/html/about2.shtml), NOAA/ National Weather Service, National Centers for Environmental Prediction, Hydrometeorological Prediction Center (http://www.hpc.ncep.noaa.gov/), Camp Springs, Maryland, 2007.

[82] "Modern research in nuclear meteorology" (http://www.springerlink.com/index/R55V691512754216.pdf) (PDF). *Atomic Energy*. Springer New York. February 1974. doi:10.1007/BF01117823. . Retrieved July 6, 2008.

Further reading

- Byers, Horace. General Meteorology. New York: McGraw-Hill, 1994.
- Garret, J.R. (1992) [1992]. *The atmospheric boundary layer*. Cambridge University Press. ISBN 0-521-38052-9.
- *Glossary of Meteorology* (http://amsglossary.allenpress.com/glossary). American Meteorological Society (2nd ed.). Allen Press. 2000.
- Bluestein, H (1992) [1992]. *Synoptic-Dynamic Meteorology in Midlatitudes: Principles of Kinematics and Dynamics, Vol. 1*. Oxford University Press. ISBN 0-19-506267-1.
- Bluestein, H (1993) [1993]. *Synoptic-Dynamic Meteorology in Midlatitudes: Volume II: Observations and Theory of Weather Systems*. Oxford University Press. ISBN 0-19-506268-X.
- Reynolds, R (2005) [2005]. *Guide to Weather*. Buffalo, New York: Firefly Books Inc. pp. 208. ISBN 1-55407-110-0.
- Holton, J.R. (2004) [2004]. *An Introduction to Dynamic Meteorology* (http://elsevier.com.uk) (4th ed.). Burlington, Md: Elsevier Inc.. ISBN 0-12-354015-1.

External links

Please see weather forecasting for weather forecast sites.

- Air Quality Meteorology (http://www.shodor.org/metweb/) - Online course that introduces the basic concepts of meteorology and air quality necessary to understand meteorological computer models. Written at a bachelor's degree level.
- The GLOBE Program (http://www.globe.gov/globe_flash.html) - (Global Learning and Observations to Benefit the Environment) An international environmental science and education program that links students, teachers, and the scientific research community in an effort to learn more about the environment through student data collection and observation.
- Glossary of Meteorology (http://amsglossary.allenpress.com/glossary) - From the American Meteorological Society, an excellent reference of nomenclature, equations, and concepts for the more advanced reader.
- JetStream - An Online School for Weather (http://www.srh.noaa.gov/srh/jetstream/) - National Weather Service
- Learn About Meteorology (http://www.bom.gov.au/lam/) - Australian Bureau of Meteorology
- The Weather Guide (http://weather.about.com) - Weather Tutorials and News at About.com
- Meteorology Education and Training (MetEd) (http://meted.ucar.edu/) - The COMET Program
- NOAA Central Library (http://www.lib.noaa.gov/) - National Oceanic & Atmospheric Administration
- The World Weather 2010 Project (http://ww2010.atmos.uiuc.edu) The University of Illinois at Urbana-Champaign
- NOAA Weather Navigator (http://dapper.pmel.noaa.gov/dchart/index.html?dsetid=e9f4fb6cf715cddaf101a13e3ce1ce9) Plot and download archived data from thousands of worldwide weather stations
- Ogimet - online data from meteorological stations of the world, obtained through NOAA free services (http://www.ogimet.com/index.phtml.en)
- Solar Eclipse Meteorological Measurement (http://hvezdarna.plzen-city.cz/zatmeni/semm/en/index.html)
- National Center for Atmospheric Research Archives, documents the history of meteorology (https://www.archives.ucar.edu/)
- Weather forecasting and Climate science (http://www.metoffice.gov.uk/learning/science) - United Kingdom Meteorological Office

Links to other keywords in meteorology	**Atmospheric conditions:** Absolute stable air	Temperature inversion	Dine's compensation	precipitation	Cyclone	anticyclone	Thermal	Tropical cyclone (hurricane or typhoon)	Vertical draft	Extratropical cyclone **Weather forecasting**: atmospheric pressure	Low pressure area	High pressure area	dew point	weather front	jet stream	wind chill	heat index	Theta-e	primitive equations	Pilot Reports **Storm**: thunderstorm	lightning	thunder	hail	tornado	convection	blizzard	supercell **Climate**: El Niño	monsoon	flood	drought	Global warming	Effect of sun angle on climate. **Air Pollution**: Air pollution dispersion modeling	List of atmospheric dispersion models	Smog **Other phenomena:** deposition	dust devil	fog	tide	wind	cloud	air mass	evaporation	sublimation	ice	crepuscular rays	anticrepuscular rays **Weather-related disasters:** weather disasters	extreme weather	list of severe weather phenomena **Climatic or Atmospheric Patterns:** Alberta clipper	El Niño	Derecho	Gulf Stream	La Niña	Jet stream	North Atlantic Oscillation	Madden-Julian oscillation	Pacific decadal oscillation	Pineapple Express	Sirocco	Siberian Express	Walker circulation

Space Weather Prediction Center

The **Space Weather Prediction Center** (**SWPC**), formerly the Space Environment Center (SEC), is a laboratory and service center of the National Oceanic and Atmospheric Administration (NOAA) National Weather Service (NWS) located in Boulder, Colorado. SWPC continually monitors and forecasts Earth's space environment, providing solar-terrestrial information. SWPC is the official source of space weather alerts and warnings for the United States.

The Space Weather Prediction Center is one of the nine National Centers for Environmental Prediction (NCEP) and provides real-time monitoring and forecasting of solar and geophysical events, conducts research in solar-terrestrial physics, and develops techniques for forecasting solar and geophysical disturbances. The SWPC Forecast Center is jointly operated by NOAA and the U.S. Air Force and is the national and world warning center for disturbances that can affect people and equipment working in the space environment. SWPC works with many national and international partners who contribute data and observations.

A few of the agencies and industry that rely on SWPC services:

- U.S. power grid infrastructure
- Commercial airline industry
- Department of Transportation (use of GPS)
- NASA human space flight activities (NASA relies on SWPC data to protect the $1 billion arm on the International Space Station)
- Satellite launch and operations
- U.S. Air Force operational support
- Commercial and public users (more than half a million hits per day on SWPC web sites)
- Some Economic Impacts of Space Weather:

The Federal Aviation Administration requires dispatchers to take into consideration HF communication degradation for each dispatched polar flight. Flights can be diverted based on SWPC solar radiation alerts if air traffic control communication is compromised, with estimated costs as high as $100K per flight. A 23-day period in 2001 saw 25 flights diverted due to such radio blackouts.

See also

- Spaceflight Meteorology Group (SMG)
- Heliophysics

External links

- Space Weather Prediction Center [1]

References

[1] http://www.swpc.noaa.gov

Storm Prediction Center

Storm Prediction Center	
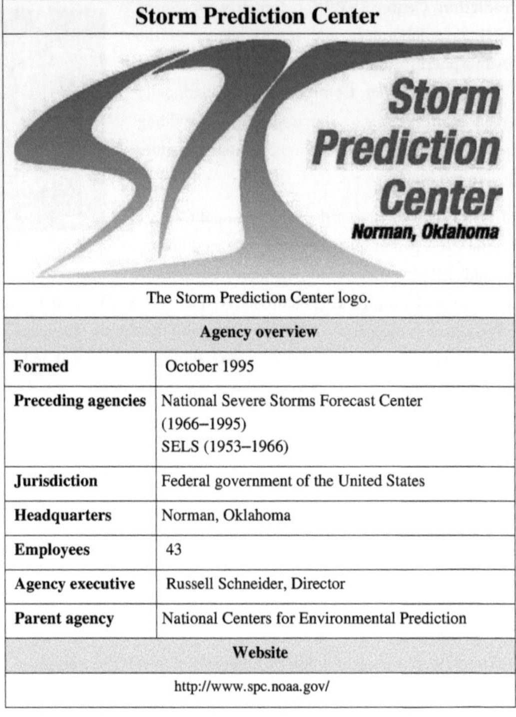	
The Storm Prediction Center logo.	
Agency overview	
Formed	October 1995
Preceding agencies	National Severe Storms Forecast Center (1966–1995) SELS (1953–1966)
Jurisdiction	Federal government of the United States
Headquarters	Norman, Oklahoma
Employees	43
Agency executive	Russell Schneider, Director
Parent agency	National Centers for Environmental Prediction
Website	
http://www.spc.noaa.gov/	

The **Storm Prediction Center** (SPC), located in Norman, Oklahoma, is tasked with forecasting the risk of severe thunderstorms and tornadoes in the contiguous United States. The agency issues convective outlooks, mesoscale discussions, and watches as a part of this process. Convective outlooks are issued for Day 1, Day 2, Day 3, and Day 4–8, and detail the risk of severe thunderstorms and tornadoes during the given forecast period, although tornado, hail, and wind details are only available for Day 1. Days 2 and 3, as well as 4–8 used a probabilistic scale, determining the probability for a severe weather event in percent. Mesoscale discussions are issued to give information on a region that is becoming a severe weather threat and states whether a watch is likely and details thereof, as well as situations of isolated severe weather when watches are not necessary. Watches are issued when forecasters are confident that severe weather will occur, and usually precede the onset of severe weather by one hour.

The agency is also responsible for forecasting fire weather (conditions favorable for wildfires) in the contiguous US, and issues Day 1, 2, and 3–8 fire weather outlooks. These outlooks detail areas with critical or extremely critical fire conditions.

The Storm Prediction Center is part of the National Centers for Environmental Prediction (NCEP), operating under the control of the National Weather Service (NWS), which in turn is part of the National Oceanic and Atmospheric Administration (NOAA) of the United States Department of Commerce (DoC).

The Storm Prediction Center was previously known as the National Severe Storms Forecast Center and was located in Kansas City, Missouri. In October 1995, the National Severe Storms Forecast Center relocated to Norman and was renamed the Storm Prediction Center. From the time of the move until 2006, it was co-located with the National

Severe Storms Laboratory at University of Oklahoma Westheimer Airport. In 2006, they moved into the National Weather Center.

History

The Storm Prediction Center began in 1952 in Washington, D.C. as SELS (**Se**vere **L**ocal **S**torms Unit), a special unit of forecasters in the Weather Bureau. In 1954, the unit moved to Kansas City. SELS began issuing convective outlooks in 1955, and began issuing radar summaries every three hours in 1960;[1] with the increased duties of radar summaries this unit became the National Severe Storms Forecast Center (NSSFC) in 1966.[2]

In 1968 the National Severe Storms Forecast Center began issuing status reports on watches, and in 1971 the agency made their first computerized data transmission.[1] On April 2, 1982 the first *particularly dangerous situation* watch was issued.[1] Two new products were introduced in 1986: the Day 2 Convective Outlook and the Mesoscale Discussion.[1]

The National Severe Storms Forecast Center remained located in Kansas City, Missouri until October 1995, when it moved to Norman, Oklahoma and was renamed the Storm Prediction Center.[3] In 1998, the Center began issuing the National Fire Weather Outlook.[1] The Day 3 Convective Outlook was first issued on an experimental basis in 2000, and was made an official product in 2001.[1] From 1995 to 2006 the Storm Prediction Center was housed at University of Oklahoma Westheimer Airport,[3] in the same building as the National Severe Storms Laboratory, after which it moved to the National Weather Center.[1]

The Storm Prediction Center continues operations out of the National Weather Center building as of 2011.[4]

Overview

The Storm Prediction Center is responsible for forecasting the risk of severe weather caused by severe thunderstorms, specifically those producing tornadoes, hail 1 inch (2.5 cm) or larger, and winds 58 mph (93 km/h) or greater. The agency also forecasts hazardous winter and fire weather. It does so primarily by issuing convective outlooks, severe thunderstorm watches, tornado watches, and mesoscale discussions.[5]

There is a three-stage process in which the area, time period, and details of a severe weather forecast are refined from a broad-scale forecast of potential hazards to a more specific and detailed forecast of what hazards are expected, where they are expected to occur, and in what time frame. If warranted, forecasts will also increase in severity through this three-stage process.[5]

The Storm Prediction Center employs a total of 43 personnel, including five lead forecasters, ten mesoscale/outlook forecasters, and seven assistant mesoscale forecasters.[6]

The Storm Prediction Center is part of the National Centers for Environmental Prediction (NCEP), operating under the control of the National Weather Service (NWS),[3] which in turn is part of the National Oceanic and Atmospheric Administration (NOAA) of the United States Department of Commerce (DoC).[7]

Many SPC forecasters and support staff are heavily involved in scientific research into severe and hazardous weather. This involves conducting applied research and writing technical papers, developing training materials, giving seminars and other presentations locally and nationwide, attending scientific conferences, and participating in weather experiments.[8]

Convective outlooks

The Storm Prediction Center issues categorical and probability forecasts describing the general threat of severe convective storms over the contiguous United States for the next 6–192 hours (Day 1–Day 8). They are labeled and issued by day and are issued up to five times a day.[9]

The categorical risks are general thunderstorms (green shaded area/previously brown line before April 2011), "SEE TEXT" (black, textual label on map indicating potential for isolated severe storms or near-severe storms), "SLGT" (yellow shaded area/previously green line indicating slight risk of severe weather), "MDT" (red shaded area/previously red line indicating moderate risk of severe weather), and "HIGH" (pink shaded area/previously fuchsia line indicating high risk of severe weather). Significant severe areas (referred to as "hatched areas" because of their representation on outlook maps) refer to a threat of increased storm intensity that is of "significant severe" (F2/EF2 or stronger tornado, 2 inches (5 cm) or larger hail, or 75 mph (120 km/h) winds or greater) level.[10]

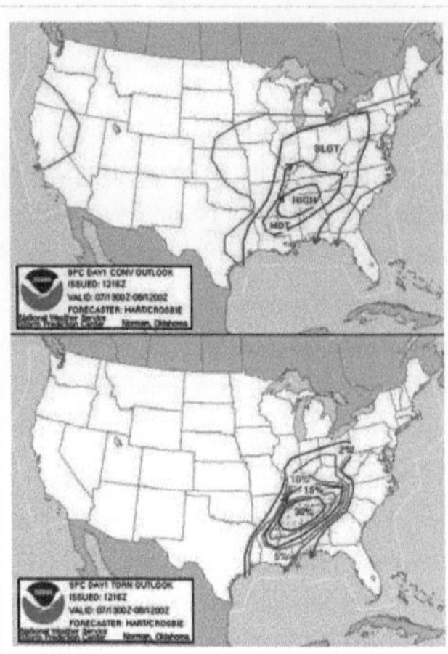

Day 1 Convective Outlook and Probabilistic maps issued by the Storm Prediction Center during the heart of a tornado outbreak on April 7, 2006. The top map indicates the risk of general severe weather (including large hail, damaging winds, and tornadoes), while the bottom map specifically shows the percent risk of a tornado forming within 25 miles (40 km) of any point within the enclosed area. The hatched area on the bottom map indicates a 10% or greater risk of an F2 or stronger tornado forming within 25 miles (40 km) of a point.

In April of 2011, the SPC began issuing new graphics for categorical and probabilistic outlooks. The new format includes shading of risk areas and population, county, and interstate overlays. The colors were changed as mentioned above as well. The new shaded maps also include changes to the probability color shades on each outlook.

Public severe weather outlooks (PWO) are issued when a significant or widespread outbreak is expected, especially for tornadoes. From November to March, it can also be issued for any threat of significant tornadoes in the nighttime hours, noting the lower awareness and greater danger of tornadoes at that time of year.[11]

Categories

A **slight risk** day typically will mean the threat exists for scattered severe weather, including scattered wind damage or severe hail and possibly some isolated tornadoes. During the peak severe weather season, most days will have a slight risk somewhere in the US. Isolated significant severe events are possible in some circumstances, but are generally not widespread.[9]

A **moderate risk** day indicates that more widespread and/or more dangerous severe weather is possible (sometimes with major hurricanes), with significant severe weather often more likely. Numerous tornadoes (including some strong tornadoes), more widespread or severe wind damage and very large/destructive hail could occur. Major events, such as large tornado outbreaks, are sometimes also possible on moderate risk days, but with greater uncertainty. Moderate risk days are not uncommon and typically occur several times a month, especially during peak season. A slight risk area typically surrounds a moderate risk area, where the threat is lower.[9]

A **high risk** day indicates a considerable likelihood of a major tornado outbreak or (much less often) an extreme derecho event. On these days, the potential exists for extremely severe and life-threatening weather, including widespread strong or violent tornadoes and/or very destructive straight-line winds (Hail cannot verify or produce a high risk on its own, although such a day usually involves a threat for widespread very large hail as well). Many of the most prolific severe weather days were high risk days. Such days are quite rare; a high risk is typically issued only a few times each year (see List of SPC High Risk days). High risk areas are usually surrounded by a larger moderate risk area, where uncertainty is greater or the threat is somewhat lower.[9]

Issuance and usage

Note: SIGNIFICANT SEVERE area needed where denoted by ***bold italic*** type – otherwise default to next lower category.

Day 1 probability to categorical outlook conversion[10]

Outlook probability	TORN	WIND	HAIL
2%	*see text*	not used	not used
5%	SLGT	*see text*	*see text*
10%	SLGT	not used	not used
15%	MDT	SLGT	SLGT
30%	HIGH	SLGT	SLGT
45%	HIGH	MDT	***MDT***
60%	HIGH	***HIGH***	MDT

Day 2 probability to categorical outlook conversion[10]

Outlook probability	Combined TORN, WIND, and HAIL
5%	*see text*
15%	SLGT
30%	SLGT
45%	***MDT***
60%	***HIGH***

Day 3 probability to categorical outlook conversion[10]

Outlook probability	Combined TORN, WIND, and HAIL
5%	*see text*
15%	SLGT
30%	SLGT
45%	***MDT***

Convective outlooks are issued in Zulu time (also known as UTC).[10]

The categories at right refer to the risk levels for the specific severe weather event occurring within 25 miles (40 km) of any point in the delineated region. SLGT is slight risk, meaning that well organized severe thunderstorms are expected, but low in number or coverage. MDT moderate risk, indicating that greater concentration and magnitude of severe weather is expected than would be expected in a slight risk. HIGH a high risk of severe weather, and indicates that a major severe weather outbreak is expected, with a high concentration of severe weather and enhanced risk of extremely severe weather. On high risk days the potential exists for 20 or more tornadoes (with some EF2 or stronger possible) or a extreme derecho (with winds of 80 miles per hour (130 km/h) or greater possible).[10] SEE TEXT means that there is a threat for severe weather, but the threat is not high enough to warrant a slight risk.[10]

The Day 1 Convective Outlook, issued five times per day at 0600Z (valid 1200Z that day until 1200Z the following day), 1300Z and 1630Z (the "morning updates," valid until 1200Z the next day), 2000Z (the "afternoon update," valid until 1200Z the next day), and the 0100Z (the "evening update," valid until 1200Z the following day), provides a textual forecast, map of categories and probabilities, and chart of probabilities. The Day 1 is currently the only outlook to issue probabilities specifically for tornadoes, hail, or wind. It is the most descriptive and highest accuracy outlook.[9]

Day 2 outlooks, issued twice daily at 0600Z and 1730Z, refer to tomorrow's weather (1200Z–1200Z of the next calendar day; for example a day 2 outlook issued on April 12, 2100 would be valid from 1200Z April 13, 2100 through 1200Z April 14, 2100) and include only a categorical outline, textual description, and a probability graph for severe convective storms generally. Day 2 moderate risks are fairly uncommon, and a Day 2 high risk has only been issued once (for April 7, 2006).[9]

Day 3 outlooks refer to the day after tomorrow, and include the same products (categorical outline, text description, and probability graph) as the Day 2 outlook. Higher probability forecasts are less and less likely as the forecast period increases due to lessening forecast ability farther in advance. No attempt is made to forecast general thunderstorms and a high risk is never issued that far out. Day 3 moderate risks are also quite rare; it has been used only ten times since the product became operational (most recently for April 27, 2011).[9] [12]

Day 4–8 outlooks are the longest-term official SPC Forecast Product, and often change significantly from day to day. They were an experimental product until March 22, 2007 when they became an official product. Areas are delineated in this forecast that have least a 30% chance of severe weather in the day 4–8 period (equivalent to a mid-range slight risk).[9]

Local forecast offices of the National Weather Service, radio and television stations, and emergency planners often use the forecasts to gauge the potential severe weather threats to their areas.[9]

Mesoscale discussions

Mesoscale discussions (MCDs) generally precede a tornado watch or severe thunderstorm watch, by 1–3 hours when possible.[13] Mesoscale discussions are designed to give local forecasters an update on a region where a severe weather threat is emerging and an indication of whether a watch is likely and details thereof, as well as situations of isolated severe weather when watches are not necessary.[13] MCDs contain meteorological information on what is happening and what is expected to happen in the next few hours, and forecast reasoning in regard to weather watches.[13] Mesoscale discussions are often issued to update information on watches already issued, and sometimes when one is to be canceled. Mesoscale discussions are also issued for winter weather and heavy rainfall events.[13]

Example

MESOSCALE DISCUSSION 0685

NWS STORM PREDICTION CENTER NORMAN OK

0253 PM CDT FRI MAY 04 2007

AREAS AFFECTED...WRN KS...PARTS OF WRN OK/ERN TX PNHDL

CONCERNING...SEVERE POTENTIAL...TORNADO WATCH LIKELY

VALID 041953Z - 042200Z

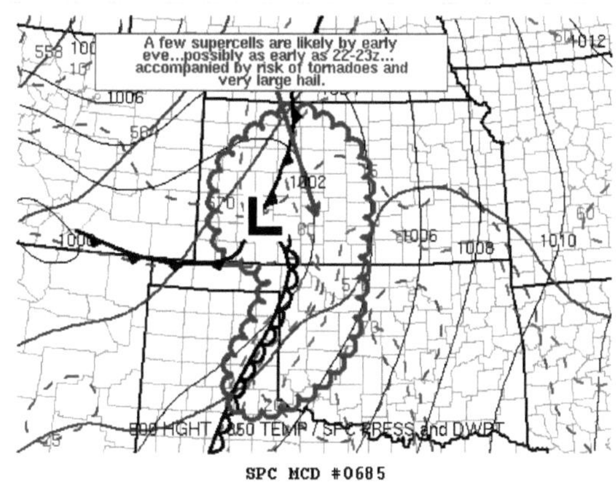

Graphic associated with the example mesoscale discussion.[14]

TRENDS ARE BEING CLOSELY MONITORED FOR SIGNS OF CONVECTIVE

INITIATION. ALTHOUGH TIMING IS STILL A BIT UNCERTAIN...ONE OR MORE

TORNADO WATCHES WILL PROBABLY BE REQUIRED LATE THIS AFTERNOON.

MOISTENING SOUTHERLY LOW-LEVEL FLOW AND STRONG DAYTIME HEATING ALONG SOUTHERN HIGH PLAINS DRY LINE...INTO THE VICINITY OF WEAK SURFACE LOW OVER SOUTHWEST KANSAS...IS CONTRIBUTING TO STRONG DESTABILIZATION. RUC GUIDANCE INDICATES MIXED LAYER CAPE IS INCREASING INTO THE 3000-4000 J/KG RANGE...THOUGH MID-LEVEL SUBSIDENCE/SHORT WAVE RIDGING ALOFT IS CURRENTLY INHIBITING CONVECTIVE DEVELOPMENT.

HOWEVER...MID/UPPER FORCING ASSOCIATED WITH AN IMPULSE LIFTING OUT OF AMPLIFIED WESTERN TROUGH IS BEGINNING TO SHIFT EAST OF THE CENTRAL/SOUTHERN ROCKIES. AND...LATEST RUC GUIDANCE SUGGESTS WEAK LOWER/MID TROPOSPHERIC COOLING HAS OCCURRED ACROSS WESTERN KANSAS INTO THE TEXAS PANHANDLE ASSOCIATED WITH IMPULSE ALREADY LIFTING NORTHWARD THROUGH THE NORTH CENTRAL HIGH PLAINS. ALTHOUGH UNCERTAINTY DOES EXISTS CONCERNING TIMING OF CONVECTIVE INITIATION...MUCH OF MODEL GUIDANCE SUGGESTS THAT THIS COULD OCCUR AS EARLY AS 22-23Z. THIS SEEMS MOST PROBABLE WHERE FORCING WILL BE STRONGEST NEAR SURFACE LOW...BUT INITIATION OF WIDELY SCATTERED STORMS MAY QUICKLY FOLLOW SUIT.

STORM DEVELOPMENT/INTENSIFICATION WILL LIKELY BE VERY RAPID ONCE CAP BREAKS. AND...LARGE CLOCKWISE CURVED LOW-LEVEL HODOGRAPHS BENEATH 40-50 KT CYCLONIC WEST SOUTHWESTERLY 500 MB FLOW WILL BE FAVORABLE FOR TORNADOES...IN ADDITION TO THE RISK OF VERY LARGE HAIL.

ISOLATED STRONG TORNADOES ARE POSSIBLE...PARTICULARLY AS LOW-LEVEL

JET STRENGTHENS NEAR/SHORTLY AFTER 04/00-01Z.

..KERR.. 05/04/2007

ATTN...WFO...ICT...OUN...GID...DDC...GLD...LUB...AMA...

Source:[14]

Weather watches

Watches issued by the SPC are generally less than 20000–50000 square miles (52000– km^2) in area and are normally preceded by a mesoscale discussion.[15] Watches are intended to be issued preceding arrival of severe weather by 1–6 hours.[15] They indicate that conditions are favorable for severe thunderstorms or tornadoes. In the case of severe thunderstorm watches organized severe thunderstorms are expected but conditions are not thought to be especially favorable for tornadoes, whereas for tornado watches conditions are thought favorable for severe thunderstorms producing tornadoes.[15] In situations where a forecaster expects a significant threat of extremely severe and life-threatening weather, a watch with special wording of "particularly dangerous situation" (PDS) is subjectively issued.[16] It is occasionally issued with tornado watches, normally for the potential of major tornado outbreaks.[16] A PDS severe thunderstorm watch is very rare and usually issued for the potential of major derecho events.[16]

Watches are not 'Warnings', where there is an immediate severe weather threat to life and property. Although Severe Thunderstorm and Tornado Warnings are ideally the next step after watches, watches cover a threat of organized severe thunderstorms over a larger area and may not always precede a warning. Warnings are issued by local National Weather Service offices, not the Storm Prediction Center, which is a national guidance center.[15]

Watches are canceled by the local National Weather Service office.[15]

Example

URGENT — IMMEDIATE BROADCAST REQUESTED

TORNADO WATCH NUMBER 232

NWS STORM PREDICTION CENTER NORMAN OK

955 AM CDT SAT MAY 5 2007

THE NWS STORM PREDICTION CENTER HAS ISSUED A

TORNADO WATCH FOR PORTIONS OF

PARTS OF WESTERN AND CENTRAL KANSAS

PARTS OF SOUTHWEST AND CENTRAL NEBRASKA

EFFECTIVE THIS SATURDAY MORNING AND EVENING FROM 955 AM UNTIL

1000 PM CDT.

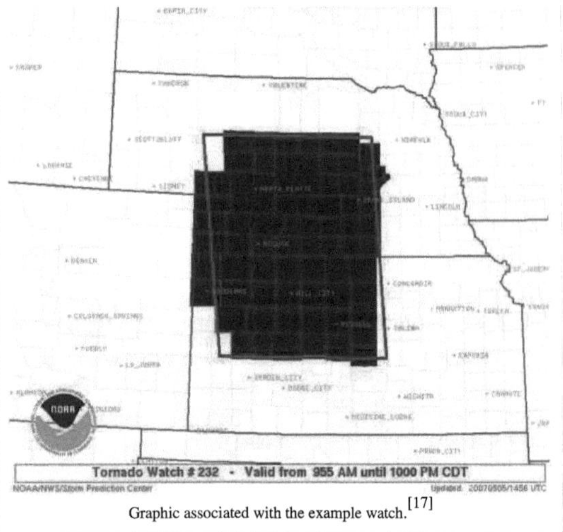

Graphic associated with the example watch.[17]

DESTRUCTIVE TORNADOES...LARGE HAIL TO 4 INCHES IN DIAMETER...

THUNDERSTORM WIND GUSTS TO 90 MPH...AND DANGEROUS LIGHTNING ARE

POSSIBLE IN THESE AREAS.

THE TORNADO WATCH AREA IS APPROXIMATELY ALONG AND 100 STATUTE

MILES EAST AND WEST OF A LINE FROM 45 MILES NORTH NORTHWEST OF

BROKEN BOW NEBRASKA TO 55 MILES SOUTHWEST OF RUSSELL KANSAS. FOR

A COMPLETE DEPICTION OF THE WATCH SEE THE ASSOCIATED WATCH

OUTLINE UPDATE (WOUS64 KWNS WOU2).

REMEMBER...A TORNADO WATCH MEANS CONDITIONS ARE FAVORABLE FOR

TORNADOES AND SEVERE THUNDERSTORMS IN AND CLOSE TO THE WATCH

AREA. PERSONS IN THESE AREAS SHOULD BE ON THE LOOKOUT FOR

THREATENING WEATHER CONDITIONS AND LISTEN FOR LATER STATEMENTS

AND POSSIBLE WARNINGS.

DISCUSSION...VERY POTENT TORNADIC SUPERCELL PATTERN IN PLACE ACROSS

WATCH AREA AS AIR MASS IS EXTREMELY UNSTABLE WITH VERY FAVORABLE

SHEAR PROFILES. WITH LITTLE INHIBITION REMAINING ALONG E OF DRY

LINE...STORMS WILL RAPIDLY BECOME SEVERE BY EARLY THIS AFTERNOON WRN

KS INTO SWRN NEB. TORNADIC SUPERCELLS WILL DEVELOP WITH POTENTIAL

FOR LONG TRACK/VIOLENT TORNADOS. AS DRY LINE REMAINS WRN KS THRU

THE AFTERNOON...ADDITIONAL DEVELOPMENT OF TORNADIC SUPERCELLS ARE

LIKELY OFF THE DRY LINE THRU THE EVENING HOURS.

AVIATION...TORNADOES AND A FEW SEVERE THUNDERSTORMS WITH HAIL

SURFACE AND ALOFT TO 4 INCHES. EXTREME TURBULENCE AND SURFACE

WIND GUSTS TO 80 KNOTS. A FEW CUMULONIMBI WITH MAXIMUM TOPS TO

600. MEAN STORM MOTION VECTOR 22040.

...HALES

Source:[17]

Fire weather products

The Storm Prediction Center also is responsible for issuing fire weather outlooks for the continental United States. These outlooks are a guidance product for local, state, and federal government agencies, including local National Weather Service offices, in forecasting the potential for wildfires.[18] The outlooks issued are for day 1, day 2, and days 3–8. The day 1 product is issued at 4:00 a.m. Central time and is updated at 1700Z, and is valid from 1200Z to 1200Z the following day. The day 2 outlook is issued at 1000Z and is updated at 2000Z for the forecast period of 1200Z to 1200Z the following day. The day 3–8 outlook is issued at 2200Z and is valid for days 3–8.[18]

An example of a Day 1 fire outlook, issued in the midst of the October 2007 California wildfires.

There are four types of Fire Weather Outlook areas: "See Text", a "Critical Fire Weather Area for Wind and Relative Humidity", an "Extremely Critical Fire Weather Area for Extreme Conditions of Wind and Relative Humidity", and a "Critical Fire Weather Area for Dry Thunderstorms".[19] The outlook type depends on forecast weather conditions, severity of the predicted threat, and local climatology of a forecast region.[18] "See Text" is a label on the map for pointing out areas where fire potential is great enough to pose a limited threat, but not enough to warrant a critical area, similar to "See Text" areas in convective outlooks. Critical Fire Weather Areas for Wind and Relative Humidity are typically issued when strong winds (>20 mph) and low Relative Humidity are expected to occur where dried fuels exist, similar to a slight or moderate risk of severe weather. Critical Fire Weather Areas for Dry Thunderstorms are typically issued when widespread or numerous thunderstorms producing little wetting rain (<0.10 in) are expected to occur where dried fuels exist. Extremely Critical Fire Weather Areas for Wind and Relative Humidity are issued when very strong winds and very low RH are expected to occur with very dry fuels. Extremely Critical areas are rarely issued, similar to the very low frequency of high risk areas in convective outlooks.[19]

See also

- Severe weather terminology (United States)
- List of SPC High Risk days

References

[1] Edwards, Roger; Fred Ostby (2009). "Timeline of SELS and SPC" (http://www.spc.noaa.gov/history/timeline.html). Storm Prediction Center. . Retrieved 2010-02-02.

[2] Stephen F. Corfidi (August 1999). "The Birth and Early Years of the Storm Prediction Center" (http://ams.allenpress.com/perlserv/?request=get-document&doi=10.1175/1520-0434(1999)014<0507:TBAEYO>2.0.CO;2). *Weather and Forecasting* (American Meteorological Society) **14** (4): 507–525. Bibcode 1999WtFor..14..507C. doi:10.1175/1520-0434(1999)014<0507:TBAEYO>2.0.CO;2. ISSN 1520-0434. .

[3] Stephen F. Corfidi (2009-12-27). "A brief history of the Storm Prediction Center" (http://www.spc.noaa.gov/history/early.html). *Storm Prediction Center*. National Oceanic and Atmospheric Administration. . Retrieved 2010-01-31.

[4] Carbin, Greg; Roger Edwards, Greg Grosshans, David Imy, Mike Kay, Jay Liang, Joe Schaefer, Rich Thompson. "Frequently Asked Questions (FAQ)" (http://www.spc.noaa.gov/faq/). *Storm Prediction Center Frequently Asked Questions*. Storm Prediction Center. . Retrieved 2010-05-13.

[5] Storm Prediction Center. "The Severe Storms Forecast Process: Outlook to Mesoscale Discussion to Watch to Warning" (http://www.spc.noaa.gov/misc/aboutus.html). *About the Storm Prediction Center*. Storm Prediction Center. . Retrieved 2009-12-27.

[6] "Storm Prediction Center Employees" (http://www.spc.noaa.gov/staff/). *spc.noaa.gov*. Storm Prediction Center. . Retrieved 2010-04-08.

[7] "NOAA's National Weather Service" (http://www.weather.gov). *weather.gov*. National Weather Service. . Retrieved 2010-05-13.

[8] *This article incorporates public domain material* (http://www.weather.gov/disclaimer.php) *from the Storm Prediction Center* document "About the Storm Prediction Center" (http://www.spc.noaa.gov/misc/aboutus.html).

[9] Novy, Chris; Roger Edwards, David Imy, Stephen Goss (2008-11-13). "Convective Outlooks" (http://www.spc.noaa.gov/misc/about.html#Convective Outlooks). *Storm Prediction Center and its Products*. Storm Prediction Center. . Retrieved 2009-12-27.

[10] Storm Prediction Center (2006-02-14). "Storm Prediction Center Day 1, 2 and 3 Convective Outlooks" (http://www.spc.noaa.gov/misc/SPC_Prob_Conv_Otlk_Change_20060214.html). National Weather Service. . Retrieved 2010-01-31.

[11] National Weather Service (2009-06-25). "Public Severe Weather Outlook" (http://www.nws.noaa.gov/glossary/index.php?word=Public+Severe+Weather+Outlook). *Glossary — National Oceanic and Atmospheric Administration's National Weather Service*. National Weather Service. . Retrieved 2010-01-31.

[12] http://www.crh.noaa.gov/news/display_cmsstory.php?wfo=lot&storyid=66428&source=0

[13] Novy, Chris; Roger Edwards, David Imy, Stephen Goss (2008-11-13). "Mesoscale Discussions" (http://www.spc.noaa.gov/misc/about.html#Convective Outlooks). *Storm Prediction Center and its Products*. Storm Prediction Center. . Retrieved 2009-12-27.

[14] Kerr, Brynn (2007-05-04). "Storm Prediction Center Mesoscale Discussion 685" (http://www.spc.noaa.gov/products/md/2007/md0685.html). *Mesoscale Discussion 685*. Storm Prediction Center. . Retrieved 2009-12-27.

[15] Novy, Chris; Roger Edwards, David Imy, Stephen Goss (2008-11-13). "Severe Weather Watches" (http://www.spc.noaa.gov/misc/about.html#Convective Outlooks). *Storm Prediction Center and its Products*. Storm Prediction Center. . Retrieved 2009-12-27.

[16] National Weather Service (2009-06-25). "PDS" (http://www.nws.noaa.gov/glossary/index.php?word=PDS). *Glossary — National Oceanic and Atmospheric Administration's National Weather Service*. National Weather Service. . Retrieved 2009-12-27.

[17] Hales, Jack (2007-05-05). "Storm Prediction Center PDS Tornado Watch 232" (http://www.spc.noaa.gov/products/watch/2007/ww0232.html). *PDS Tornado Watch 232*. Storm Prediction Center. . Retrieved 2009-12-27.

[18] Novy, Chris; Roger Edwards, David Imy, Stephen Goss (2010-03-25). "Fire Weather Outlooks" (http://www.spc.noaa.gov/misc/about.html#Convective Outlooks). *Storm Prediction Center and its Products*. Storm Prediction Center. . Retrieved 2010-04-16.

[19] *This article incorporates public domain material* (http://www.weather.gov/disclaimer.php) *from the Storm Prediction Center* document "Fire weather outlooks" (http://www.spc.noaa.gov/misc/about.html#FireWx).

External links

- Storm Prediction Center (SPC homepage) (http://www.spc.noaa.gov)
- SPC products descriptions (http://www.spc.noaa.gov/misc/about.html)

National Hurricane Center

National Hurricane Center	
Agency overview	
Formed	1967
Jurisdiction	United States government
Headquarters	Miami, Florida
Agency executives	Bill Read, Director Edward Rappaport, Deputy Director
Website	
[1]	

The **National Hurricane Center** (**NHC**), located at Florida International University in Miami, Florida, is the division of the National Weather Service responsible for tracking and predicting weather systems within the tropics between the Prime Meridian and the 140th meridian west poleward to the 30th parallel north in the northeast Pacific ocean and the 31st parallel north in the northern Atlantic ocean. Its Tropical Analysis and Forecast Branch (TAFB) routinely issues marine forecasts, in the form of graphics and high seas forecasts, for this area year round.

During the Atlantic and northeast Pacific hurricane seasons, the Hurricane Specialists Unit issues routine tropical weather outlooks for the northeast Pacific and northern Atlantic oceans. When tropical storm or hurricane conditions are expected within 36 hours, the center issues the appropriate watches and warnings via the news media and NOAA Weather Radio. Although the NHC is an agency of the United States, the World Meteorological Organization has designated it as Regional Specialized Meteorological Center for the North Atlantic and eastern Pacific. As such, the NHC is the central clearinghouse for all tropical cyclone forecasts and observations occurring in these areas, regardless of their effect on the US. If the center loses power or becomes incapacitated in some manner, the Central Pacific Hurricane Center backs tropical cyclone advisories and tropical weather outlooks for the northeast Pacific ocean while the Hydrometeorological Prediction Center backs up tropical cyclone advisories and tropical weather outlooks for the north Atlantic ocean.

History

The National Hurricane Center has its roots in a December 19, 1898 declaration by then-President William McKinley for the Weather Bureau (now the National Weather Service) to establish a hurricane warning network. As communications and forecasting evolved, responsibility for issuing hurricane warnings was eventually centralized in the Miami Weather Bureau office.[2]

The Miami office was designated the National Hurricane Center in 1967, and given responsibility for Atlantic tropical cyclones in their vicinity. Some other hurricane warning centers, such as in New Orleans and Boston, played a role even into the 1980s. In 1984, the NHC was separated from the Miami Weather Service Forecast Office, which meant the meteorologist in charge at Miami was no longer in a position above the hurricane center director. By 1988, the NHC gained responsibility for eastern Pacific tropical cyclones as the former Eastern Pacific Hurricane Center in San Francisco was decommissioned.

In 1992, Hurricane Andrew blew the WSR-57 weather radar and the anemometer off the roof of Gables One Tower, then the location of the NHC's offices. The radar was replaced with a WSR-88D NEXRAD system. In 1995, the NHC moved into a new hurricane resistant facility on the campus of Florida International University, capable of withstanding 130 mph (210 km/h) winds.[3] The current director of the National Hurricane Center is Bill Read.

Former and current directors

- Gordon Dunn (1965–1967)
- Robert Simpson (1967–1973), co-creator of the Saffir-Simpson Hurricane Scale.
- Neil Frank (1973–1987), responsible for creating strong ties between NHC and the media while director.
- Bob Sheets (1987–1995)
- Bob Burpee (1995–1997)
- Jerry Jarrell (1998–2000)
- Britt Max Mayfield (2000–2007)
- Xavier William Proenza (2007)[4]
- Edward (Ed) N. Rappaport (2007–2008)
- Bill Read (2008–Current)

Tropical Analysis and Forecast Branch

A panoramic view of TAFB's operations at the NHC

The **Tropical Analysis and Forecast Branch** (**TAFB**, formerly the **Tropical Satellite Analysis and Forecast unit**) is a part of the National Hurricane Center in Miami, Florida. The TAFB is responsible for high seas analyses and forecasts for tropical portions of the Atlantic and Pacific. Unlike the Hurricane Specialists Unit (HSU), TAFB is staffed full-time around the year. Other responsibilities of the TAFB include satellite-derived tropical cyclone position and intensity estimates, WSR-88D radar fixes for tropical cyclones, tropical cyclone forecast support, media support, and general operational support.[5]

Hurricane specialists

The NHC's hurricane specialists are the chief meteorologists that predict the actions of tropical storms. The specialists work rotating eight-hour shifts from May through November, monitoring weather patterns in the Atlantic and Eastern Pacific oceans. Whenever a depression appears, they issue advisories every six hours until the storm runs its course. Public advisories are issued more often when the storm threatens land. The specialists coordinate with officials in each country likely to be affected. They forecast and recommend watches and warnings.

Each specialist signs forecasts and advisories with their last name, sometimes issuing joint statements with other NHC staff members.

Outside of the hurricane season, the specialists concentrate on public education efforts.[6] [7]

Hurricane naming process

In the 1953 Atlantic season, the United States Weather Bureau began naming storms which reach tropical storm intensity with human names. This replaced a 3-year plan (involving the 1950, 1951, and 1952 hurricane seasons) to name storms using the spelling alphabet.[8] Initially, storms only had female names, but after some protest, male and female names were alternated beginning in the 1979 season.[9]

The World Meteorological Organization now creates and maintains the annual lists. Names are used on a six-year rotation, with the deadliest or most notable storms having their names retired from the rotation.

Research

As part of their annual tropical cyclone activity, the agency issues a tropical cyclone report on every tropical cyclone in the Atlantic and Eastern Pacific Ocean basins, which are available since 1958 and 1988, respectively. The report summarizes the synoptic history, meteorological statistics, casualties and damages, and the post-analysis best track of a storm.[10] The reports were formally known as Preliminary Reports up until 1999.[11]

The agency maintains archives and climatological statistics on Atlantic and Pacific hurricane history, including annual reports on every tropical cyclone, a complete set of tropical cyclone advisories, digitized copies of related materials on older storms, season summaries published as the Monthly Weather Review, and HURDAT, which is the official tropical cyclone database.[12]

See also

- Canadian Hurricane Centre, an Tropical Cyclone Warning Center within NHC's area of responsibility
- Central Pacific Hurricane Center
- Joint Typhoon Warning Center
- George R. Stewart
- Storm (novel)

References

[1] http://www.nhc.noaa.gov/index.shtml

[2] American Presidency Project. William McKinley. (http://www.presidency.ucsb.edu/ws/index.php?pid=29539) Retrieved on 6 December 2006.

[3] Ametsoc.org (http://www.ametsoc.org/amschaps/nov02news.html)

[4] "Commerce Secretary and NOAA Administrator announce new National Hurricane Center Director Bill Proenza to Succeed Max Mayfield." (http://www.noaanews.noaa.gov/stories2006/s2752.htm). National Oceanic and Atmospheric Administration. 6 December 2006. . Retrieved 2006-12-06.

[5] Tropical Analysis and Forecast Branch (http://www.nhc.noaa.gov/abouttafb.shtml)

[6] Tropical Prediction Center. The National Hurricane Center: Max Mayfield, Director Ed Rappaport, Deputy Director. (http://www.nhc.noaa.gov/aboutnhc.shtml) Retrieved on 6 December 2006.

[7] Tropical Prediction Center. Tropical Prediction Center Staff. (http://www.nhc.noaa.gov/aboutstaff.shtml) National Hurricane Center. Retrieved on 2009-03-08.

[8] Gary Padgett (1999). "Monthly Global Tropical Cyclone Summary July 2007" (http://www.webcitation.org/5tItxgNYc). Australian Severe Weather. Archived from the original (http://australiasevereweather.com/cyclones/2008/summ0707.htm) on 2010-10-07. . Retrieved 2010-10-07.

[9] Colin J. McAdie, Christopher W. Landsea, Charles J. Neumann, Joan E. David, Eric S. Blake, Gregory R. Hammer (2009-08-20) (PDF). *Tropical Cyclones of the North Atlantic Ocean, 1851 – 2006* (http://www.nhc.noaa.gov/pdf/TC_Book_Atl_1851-2006_lowres.pdf) (Sixth ed.). National Oceanic and Atmospheric Administration. p. 18. . Retrieved 2010-07-07.

[10] NOAA Coastal Services Center. "Historical Hurricane Tracks" (http://maps.csc.noaa.gov/hurricanes/reports.jsp). National Oceanic and Atmospheric Administration. . Retrieved 2008-11-03.

[11] National Hurricane Center (2008). "2008 Atlantic hurricane season" (http://www.nhc.noaa.gov/2008atlan.shtml). NOAA. . Retrieved 2008-11-19.

[12] National Hurricane Center staff (2011-05-10). "NHC Archive of Hurricane Seasons" (http://www.nhc.noaa.gov/pastall.shtml). National Oceanic and Atmospheric Administration. . Retrieved 2011-07-03.

External links

- NOAA National Hurricane Center website (http://www.nhc.noaa.gov/)
- History of NWS Forecast Office, Miami (http://www.srh.noaa.gov/mfl/history/)
- National Hurricane Center (http://twitter.com/atlanticwatch) on Twitter (note this is unofficial)

Climate Prediction Center

The **Climate Prediction Center** (CPC) is one of the National Centers for Environmental Prediction, which are a part of NOAA's National Weather Service. It is located in Camp Springs, Maryland.

Products

The CPC's products are operational predictions of climate variability, real-time monitoring of global climate, and attribution of the origins of major climate anomalies. The products cover time scales from a week to seasons, and cover the land, the ocean, and the atmosphere, extending into the stratosphere.

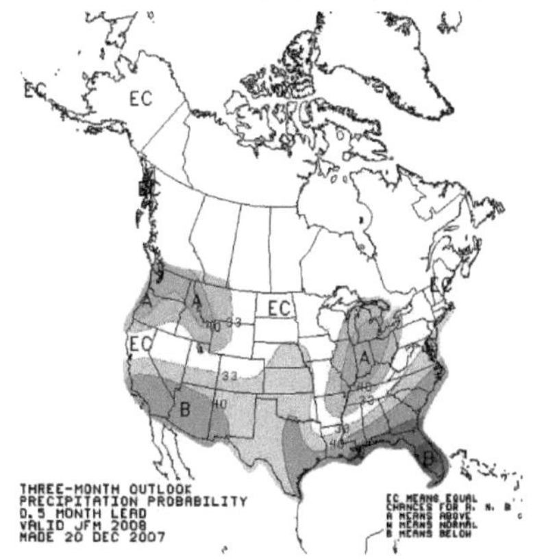

Sample CPC graphic: three month precipitation outlook

These climate services are available for users in government, the public and private industry, both in this country and abroad. Applications include the mitigation of weather related natural disasters and uses for social and economic good in agriculture, energy, transportation, water resources, and health. Continual product improvements are supported through diagnostic research, increasing use of models, and interactions with user groups. Some specific products include:

- 3-Month Temperature and Precipitation
  - Outlooks
  - Discussions
- 1-Month Temperature and Precipitation
  - Outlooks
  - Discussions
- 6 to 10-Day and 8 to 14-Day Products
  - Temperature and Precipitation Anomaly
  - Excessive Heat Outlook
  - Maximum Heat Index Prediction
- 3-Month Probability of Exceedance
  - Temperature
  - Precipitation
  - Heating and Cooling Degree Days
- Hurricane Season Outlook
  - Atlantic basin

- Pacific basin
- U.S. Drought
  - Outlook
  - Discussion
- International Support
  - Weekly Afghan Hazards
  - Weekly Africa Hazards
  - Weekly Central America Hazards
  - Weekly Haiti Hazards

History

The roots of modern climate prediction can be traced to the late 18th century. One of the nation's first applied climatologists was Thomas Jefferson, the third President of the United States. A century later, the federal government assigned to the Army Signal Corps the mission to define the climate of the regions of the country being opened for farming.

In 1890, the United States Department of Agriculture (USDA) created the Weather Bureau climate and crops services which began publishing the Weather and Crops Weekly Bulletin, which the CPC in conjunction with the USDA still publishes today.

In 1970, various federal weather and climate functions were consolidated into the National Weather Service (NWS) and placed in a new agency called the National Oceanic and Atmospheric Administration (NOAA). In the 1980s the National Weather Service established the Climate Prediction Center, known at the time as the Climate Analysis Center (CAC). The CPC is best known for its United States climate forecasts based on El Niño and La Niña conditions in the tropical Pacific.

See also

- National Climatic Data Center
- Climatology

External links

- http://www.cpc.noaa.gov/index.php

Article Sources and Contributors

Environmental Modeling Center *Source*: http://en.wikipedia.org/w/index.php?title=Environmental_Modeling_Center *Contributors*: Edward, EncycloPetey, Ks0stm, MDfoo, Martarius, PigFlu Oink, William M. Connolley, 1 anonymous edits

Data assimilation *Source*: http://en.wikipedia.org/w/index.php?title=Data_assimilation *Contributors*: Actw, Bobfry, CBM, Cmf576, Cremepuff222, Deditos, Dlary, Drf5n, Edward, Georgia1954, Grafen, Hailcloud, Ikovalets, Jmath666, Laurentdk01, Melcombe, Mkfwd, NHSavage, Netkinetic, Pegminer, Peterlean, Pflatau, Quimba, R'n'B, Radak, Rahnle, Rfinchdavis, Rnt20, RockMagnetist, Silverfish, TomyDuby, Vik-Thor, 52 anonymous edits

National Centers for Environmental Prediction *Source*: http://en.wikipedia.org/w/index.php?title=National_Centers_for_Environmental_Prediction *Contributors*: A2Kafir, Alan Liefting, Bazonka, Christopher Hollis, Cucurbitaceae, DAK4Blizzard, Evolauxia, G Clark, Jason Rees, Jasonmberry, KeelNar, Ks0stm, Lightmouse, MDfoo, Mm ncep cso, Triberocker, Wayland, William M. Connolley, 5 anonymous edits

National Oceanic and Atmospheric Administration *Source*: http://en.wikipedia.org/w/index.php?title=National_Oceanic_and_Atmospheric_Administration *Contributors*: A7x, Ademine, Ahoerstemeier, Akeane, Alan Liefting, Alansohn, Aleuts, Amillar, Andros 1337, Andy Marchbanks, Ardonik, Arthur Rubin, Badseed, Barliner, Bart133, Being blunt, Bergsten, Berkut, Between the Hammer and the Anvil, Bevo, Bob Northug, Bobo192, Bongwarrior, Brennanhay, Briaboru, CapitalR, Chemboss, Cla68, Clotho, CopperSquare, CultureDrone, Cybercobra, Cyrius, D6, Dannerswork, DeadEyeArrow, Delirium, Denelson83, Dgabbard, Eastlaw, Ed Poor, ElBenevolente, Emmiej, Epolk, Famartin, Fanghong, Farras Octara, Flyhighplato, Frenstad, G Clark, Gobonobo, Graham87, Green Giant, GrouchyDan, Hackshaven, Halftank, Hard Raspy Sci, Heidijane, Hurricanehink, Hydrargyrum, Ianwatts, Infoman99, Infrogmation, Insituburn, Instantnood, IronMaidenRocks, Ironiridis, Isomorphic, Itai, JTN, Jackthelizard, Jer10 95, Joey80, John Broughton, John222222, JonathanDP81, JonathanLamb, Jorfer, Jose Concepcion, Jpers36, Judsonfeder, Kairologic, Katenielsen2005, Keith Edkins, Kmccoy, Kozuch, LeoC4, LeonardoGregianin, Lightmouse, LilHelpa, MCalamari, Mav, Maverick9711, Mbellavia, Mbeychok, Mdnavman, Melcombe, Mellen22, Mercury34, Michael Hardy, MilFlyboy, Minesweeper, MisfitToys, MrDolomite, Muijz, Murray Langton, MyFavoriteMartin, N328KF, Neovu79, Nopetro, Oceanexplorer, PDH, PRoy1956, PaulHanson, Pauley2483, Pearle, Pgan002, Piano non troppo, Pierre cb, Poor Yorick, Postdlf, Prunesqualer, Puchiko, Quadell, RJBurkhart3, Raul654, Raven in Orbit, Redlentil, Rich Farmbrough, Rillian, Robomaeyhem, Rougher07, Ryjaz, SDC, SGT141, ST47, Sardanaphalus, Scriberius, Sdsds, Sfgiants906, Shirulashem, Siddhant, SimonP, Skew-t, Skier Dude, Sm8900, StaticGull, Surv1v4l1st, Ta bu shi da yu, Tainter, Tedernst, The Cunctator, The Nut, Tide rolls, Titoxd, Tom, TomStar81, Tommy Gao, Two Hearted River, Uaimh, Utcursch, Vicarage, Voidxor, Vrray people1000, Vsmith, W90, Waterboy12, Wavelength, William M. Connolley, Woohookitty, Wyatts, XU-engineer, Zachary, Zyxw, شرز, 139 anonymous edits

Marine Modeling and Analysis Branch *Source*: http://en.wikipedia.org/w/index.php?title=Marine_Modeling_and_Analysis_Branch *Contributors*: Addere, Alan Liefting, Emersoni, Jdorje, MDfoo, William M. Connolley, 3 anonymous edits

Climatology *Source*: http://en.wikipedia.org/w/index.php?title=Climatology *Contributors*: APH, Agnana, Ahoerstemeier, AlexD, Alfio, Alsandro, Andreas Willow, Anticipation of a New Lover's Arrival, The, Anypodetos, AstroHurricane001, Aude, Avoided, Benwildeboer, Birdbrainscan, CMD Beaker, Cedrium, Centrx, Chantal vieuille, Chevinki, Childhoodsend, Christopher Kraus, Ciaobabyx46, Ciphers, Daniel Collins, DavidH, Davidstrauss, DerHexer, Discospinster, Dragons flight, E Wing, Ed Poor, Edmeister, Edward, Eh kia, El C, Eleassar777, Emperorbma, Evolauxia, Father Goose, ForgottenLore, Gabriel Kielland, Gaius Cornelius, Gilliam, Hard Raspy Sci, Hooperbloob, Id447, Jackol, Jason Patton, Jclemens, Jpgordon, Jusjih, Kaare, Kentin, Kevin Brumage, Kinston eagle, Kummi, KurtLC, Leonidasthespartan, Lightmouse, Luna Santin, Maniwar, Maurreen, Maxim, Mayumashu, Megan Reyes, Melcombe, Michael Hardy, Movementarian, Mxn, N p holmes, Natevw, Nathan Johnson, Neilc, Netson76, Nubiatech, Obradovic Goran, PericlesofAthens, Pflatau, Piano non troppo, Poccil, Prashanthns, Prolog, Publixx, QuackGuru, Raul654, Raymond arritt, Recognizance, Rjwilmsi, Robert Merkel, SEWilco, Seaphoto, Sewing, Shiftchange, Shirulashem, Simplex1swrhs, Stephan Schulz, THEN WHO WAS PHONE?, TMLutas, TheWoodsman, Thegreatdr, Tide rolls, Tonderai, Tony1, Triberocker, Vextron, Vsmith, Waggers, Wavelength, William M. Connolley, Wolfkeeper, Wombatko, Yk Yk Yk, 125 anonymous edits

Meteorology *Source*: http://en.wikipedia.org/w/index.php?title=Meteorology *Contributors*: 13alexander, 15guzmanrubi, 2004-12-29T22:45Z, 2nd Piston Honda, 5 albert square, ABF, Acalamari, Acroterion, Adam Zábranský, Addshore, AgadaUrbanit, Ageekgal, AgentPeppermint, Ahoerstemeier, Ahunt, Ajonlime, Alansohn, AlexD, Alexwcovington, AlistairMcMillan, Altenmann, AnakngAraw, Anclation, Andrejj, Andres, Andrewmc123, Andris, Andy M. Wang, Andyjsmith, Antandrus, Anypodetos, Archeus, Ardg08, Asherandshelley, Avono, Ayrton Prost, BRG, Bachrach44, Bd64kcmo, Beaumont, Ben-Zin, Benjaminevans82, Betacommand, Bingo-101a, Blanchardb, Blehfu, BlueGoose, Bob A, Bob rulz, Bobo192, Boing! said Zebedee, Bongwarrior, BradBeattie, Brantgoose, Brent bray, CChips, CWii, Calor, Caltas, CambridgeBayWeather, Can't sleep, clown will eat me, CanadianLinuxUser, Capricorn42, CesarB, Chessie45, Chicago8, Christian75, Civil Engineer III, Ckatz, Cleonis, CloutierFan02, Cmapm, Colonies Chris, Connormah, Conrad Leviston, Conversion script, Corruptcopper, Courcelles, CrazyC83, Crimsone, Cutler, Cymru.lass, D6, DMacks, DVD R W, Danger, Darkmoon802, Darth Panda, DasianSensation, DavidH, DavidLevinson, Dawnseeker2000, Dcwinds, Dddstone, Deditos, Demonburrito, Denelson83, Denni, Deor, Dertius, Dfrg.msc, Dialectric, Diamond2, Diliff, Dinosaurdarrell, Discospinster, Dlary, Dlohcierekim, Docu, Donama, Draginator, DragonflySixtyseven, Dresdnhope, Drinking.coffee, Drmies, Dsmdgold, Duffman, EAGLESCOUT185, ESkog, EWS23, EamonnPKeane, East718, Ec5618, Ed Cormany, Egil, El C, Elassint, Eleanor Y, Eleland, Eliz81, Elkman, Elockid, Ensa, Enuja, Enviroboy, Epbr123, Erebus Morgaine, Erickroh, Esnascosta, Eugene van der Pijll, Euryalus, Evolauxia, Excirial, Faincut, Fair Deal, Fallenangel2009, Feline Hymnic, Flume, Fredbauder, Frenchwhale, FritsKoek, GLaDOS, GUllman, Gabriel Kielland, Gaius Cornelius, Geoffrey.landis, GhostofSuperslum, Giftlite, Gilliam, Gimboid13, Glenn, Gogo Dodo, Goobergunch, Graham87, Greyhood, Grim23, GrouchyDan, HRS IAM, Hadal, Halmstad, HamburgKiwi, Hard Raspy Sci, Harland1, Hdt83, Headbomb, Herakles01, Herbee, Hmhorselover, Hurricane Omega, Hyad, IRP, Iltmuw, Immunize, Insanity Incarnate, Instantnood, Inwind, Iridescent, Isopropyl, J.delanoy, J8079s, JMK, JSR, JTN, Ja 62, Jackol, Jagged 85, Janielarson, Jason Patton, Jayantanth, Jb6075, Jbergerot, Jbergste, Jd027, JdH, Jesse.enloe, Jkasd, John254, JohnOwens, JonHarder, Jpkoester1, Jrdioko, Jshook08, Juliancolton, Jusdafax, KNewman, Kajirus1, Kalathalan, Karol Langner, Keahapana, Kelisi, KennethJ, Knight1993, Knowledge Incarnate, Knutux, Koavf, Kornemuz, Kpjas, Kren46, Kwamikagami, LeaveSleaves, Leonard^Bloom, LilHelpa, Lir, Loren36, LovesMacs, Luckymama58, MK8, Mackensen, Magioladitis, Makeswell, Mani1, Manuel Trujillo Berges, Marek69, Martin Osterman, Mattbrundage, Matthew Stannard, Mattquigley, Mav, Maximillion Pegasus, Mbeychok, Mechwarrior Puppies, Mejor Los Indios, Melaen, Mendaliv, Mereda, Merovingian, Michael Hardy, MichaelBillington, Mike Rosoft, Mild Bill Hiccup, Minesweeper, MisterCharlie, Mmachon, Mmoneypenny, Morn, Msinummoc, Mujumdar, Mushin, Mwtoews, Mxn, MyFavoriteMartin, Mygerardromance, N5iln, NCAR Archives, NHSavage, Nakon, Natalie Erin, Neilc, Netoholic, Nev1, Nickhasmaya, Nicklott, Nomad2u2001, Noroton, Nsaa, Nuno Tavares, Octahedron80, Olivier, Orange Suede Sofa, Ottergoose, Paul-L, Peterlean, Pgan002, Pharaoh of the Wizards, Piano non troppo, Plange, Plasticup, Plaws, Plumbago, Pne, Poetaris, PoliticalJunkie, Polyamorph, Pparrish, Prolog, Puchiko, Pylori, Quintin3265, R'n'B, RJBurkhart3, RadarCzar, RadiantRay, Random user 39849958, RandomP, RattleMan, Raul654, Rbeas, Rd232, Reach Out to the Truth, Retaggio, RexNL, Rholton, Rich Farmbrough, Ricky81682, Rjwilmsi, Rob Hooft, Robomaeyhem, Runningonbrains, Ruy Pugliesi, SEWilco, SWAdair, Sampsonman, Samsee, Samuel, Sanya, Sardanaphalus, Satori Son, Schaengel89, SensingWater, Sensitive Oscillator, Shoessss, Short Brigade Harvester Boris, Shruti14, Siegfriedcqb, Signalhead, SimonP, Skew-t, Skomae, Skunkboy74, Skysmith, Slammus aran, Slon02, SmthManly, Spacepotato, Spellmaster, Spliffy, Stavlor, Stefeyboy, Steve (usurped), Sturtone, Sue Rangell, Sunyoswego, Sxhpb, Syncategoremata, TFD3, Tabletop, Tarquin, The Anome, The Rambling Man, The Thing That Should Not Be, The Transhumanist, TheKMan, Thedjatclubrock, Thegreatdr, Thomas Yeardly, Tiddly Tom, Tifego, Titoxd, Tobby72, Togo, Tom, Tony Sidaway, Topbanana, Trevor MacInnis, Treyt021, Triwbe, Truman1177, Twelvethirteen, Tximist, Una Smith, Urhixidur, UtherSRG, Vanished User 4517, Vary, Verrai, Versus22, Vicarious, Viggetto, Viridian, Virtualsets, Voyevoda, Vsmith, WOSlinker, Waldir, Wavelength, Weathermandan, Wellington, Wi-king, William M. Connolley, Willking1979, Willsmith, Wolfkeeper, Xcentaur, Yahia.barie, Yanksox, Yesyoudid, Youssefsan, Yuriybrisk, Yv1hx, Zedla, Zfarthin, Zigger, Zizonus, Александър, ם'אפר לורט, 787 anonymous edits

Space Weather Prediction Center *Source*: http://en.wikipedia.org/w/index.php?title=Space_Weather_Prediction_Center *Contributors*: Danthemankhan, Denelson83, Epolk, Evolauxia, HarryAlffa, Ks0stm, Leuliett, OpenToppedBus, Pearle, Pierre cb, Wayland, 1 anonymous edits

Storm Prediction Center *Source*: http://en.wikipedia.org/w/index.php?title=Storm_Prediction_Center *Contributors*: BigT27, Blairtrosper, Bollar, Caiaffa, CalebNoble, Clindberg, CrazyC83, Cyrius, Dan100, EdJogg, Evolauxia, Feline Hymnic, Gamaliel, Grey Wanderer, H1nkles, Hurricanehink, J4lambert, Jason Rees, Jezhotwells, Juliancolton, Koavf, Ks0stm, Leonard^Bloom, Muhandes, Oubrioko, Pierre cb, Radiojon, Rich257, Rjm656s, Rjwilmsi, Rptrcub, RyguyMN, Sd31415, Southern Illinois SKYWARN, Ta bu shi da yu, TexasAndroid, Tyler10M, Weatherstar4000, Wiki alf, WxGopher, Wxfan406, Xnatedawgx, 46 anonymous edits

National Hurricane Center *Source*: http://en.wikipedia.org/w/index.php?title=National_Hurricane_Center *Contributors*: A2Kafir, Ajm81, Andros 1337, Andy Marchbanks, AySz88, Bevo, Billinghurst, CQ, CallumBarney, Cdc, Chacor, Coldcaffeine, Comayagua99, Coredesat, Coreywalters06, CrazyC83, CrazyTalk, Crazyscreenwriter, Cybercobra, Cyrius, Dbaron, Delirium, Emperorbma, Francvs, Gillis54, GraemeL, GroveGuy, Gunnk, Harald Hansen, Hqb, HurricaneJeanne, HurricaneTeen, Hurricanehink, J.delanoy, JB82, Jaranda, Jason Rees, Jdorje, John W. Kennedy, Jyril, KeithH, Korranus, Ks0stm, Lars Washington, Loser user, Lotje, MDfoo, Mateus Hidalgo, Maverick9711, Miss Madeline, Moondyne, Mulad, NSLE, Namflnamfl, Nascar1996, Nationalparks, Neutrality, Philip Trueman, Plasticup, Quarty, Quidam65, Ramisses, RattleMan, Rich Farmbrough, Sarsaparilla39, Scriberius, Slysplace, Super-Magician, Ta bu shi da yu, Thatmarkguy, Thegreatdr, Thespian, Titoxd, Tom, TommyBoy, Tropische Storm Sven, Trvsdrlng, Wachholder0, Wavehunter, WindRunner, Wxweenie91, Xmort, Чръный человек, 103 anonymous edits

Climate Prediction Center *Source*: http://en.wikipedia.org/w/index.php?title=Climate_Prediction_Center *Contributors*: Gwinva, Hmains, JonathanLamb, Ks0stm, Levineps, Tidal923, Triberocker, William Avery, 4 anonymous edits

Image Sources, Licenses and Contributors

file:NOAA logo.svg *Source*: http://en.wikipedia.org/w/index.php?title=File:NOAA_logo.svg *License*: unknown *Contributors*: User:Badseed

Image:NOAAPreparing.jpg *Source*: http://en.wikipedia.org/w/index.php?title=File:NOAAPreparing.jpg *License*: unknown *Contributors*:

Image:NOAA WP-3D Orions.jpg *Source*: http://en.wikipedia.org/w/index.php?title=File:NOAA_WP-3D_Orions.jpg *License*: unknown *Contributors*: NOAA

Image:NOAA-Corplogo.jpg *Source*: http://en.wikipedia.org/w/index.php?title=File:NOAA-Corplogo.jpg *License*: unknown *Contributors*: Jamesofur, Juiced lemon, Kyle, Mattes, Superm401

Image:US-NationalWeatherService-Logo.svg *Source*: http://en.wikipedia.org/w/index.php?title=File:US-NationalWeatherService-Logo.svg *License*: unknown *Contributors*: National Oceanic and Atmospheric Administration

Image:NOAAEngineerAtWork.jpg *Source*: http://en.wikipedia.org/w/index.php?title=File:NOAAEngineerAtWork.jpg *License*: unknown *Contributors*: 1 anonymous edits

Image:NOAA Flag.svg *Source*: http://en.wikipedia.org/w/index.php?title=File:NOAA_Flag.svg *License*: unknown *Contributors*: w:en:User:MdnavmanMdnavman

Image:Annual Average Temperature Map.jpg *Source*: http://en.wikipedia.org/w/index.php?title=File:Annual_Average_Temperature_Map.jpg *License*: unknown *Contributors*: Dragons flight, Vonvon, 3 anonymous edits

Image:El Nino regional impacts.gif *Source*: http://en.wikipedia.org/w/index.php?title=File:El_Nino_regional_impacts.gif *License*: unknown *Contributors*: NOAA

Image:La Nina regional impacts.gif *Source*: http://en.wikipedia.org/w/index.php?title=File:La_Nina_regional_impacts.gif *License*: unknown *Contributors*: NOAA

Image:MJO 5-day running mean through 1 Oct 2006.png *Source*: http://en.wikipedia.org/w/index.php?title=File:MJO_5-day_running_mean_through_1_Oct_2006.png *License*: unknown *Contributors*: NOAA

File:AlpineRainbow.jpg *Source*: http://en.wikipedia.org/w/index.php?title=File:AlpineRainbow.jpg *License*: unknown *Contributors*: AgadaUrbanit at en.wikipedia

File:Baker beach at twilight 41.jpg *Source*: http://en.wikipedia.org/w/index.php?title=File:Baker_beach_at_twilight_41.jpg *License*: unknown *Contributors*: Mila Zinkova

File:Wea00920.jpg *Source*: http://en.wikipedia.org/w/index.php?title=File:Wea00920.jpg *License*: unknown *Contributors*: Original uploader was Dhaluza at en.wikipedia

File:Earth Global Circulation.jpg *Source*: http://en.wikipedia.org/w/index.php?title=File:Earth_Global_Circulation.jpg *License*: unknown *Contributors*: Joey-das-WBF, Lumijaguaari, Pierre cb, Raeky, Saperaud, Wricardoh, 2 anonymous edits

File:Wolkenstockwerke.png *Source*: http://en.wikipedia.org/w/index.php?title=File:Wolkenstockwerke.png *License*: unknown *Contributors*: Dbc334, Mr. B.B.C., Saperaud, Wst

File:Weather Bureau 1965.jpg *Source*: http://en.wikipedia.org/w/index.php?title=File:Weather_Bureau_1965.jpg *License*: unknown *Contributors*: Demonburrito

File:Huracán Hugo.jpg *Source*: http://en.wikipedia.org/w/index.php?title=File:Huracán_Hugo.jpg *License*: unknown *Contributors*: Barcex, Denniss, Rosarinagazo

File:Surface analysis.gif *Source*: http://en.wikipedia.org/w/index.php?title=File:Surface_analysis.gif *License*: unknown *Contributors*: Pierre cb, Runningonbrains, W!B:

File:WOA05 sea-surf TMP AYool.png *Source*: http://en.wikipedia.org/w/index.php?title=File:WOA05_sea-surf_TMP_AYool.png *License*: unknown *Contributors*: User:Plumbago

File:Day5pressureforecast.gif *Source*: http://en.wikipedia.org/w/index.php?title=File:Day5pressureforecast.gif *License*: unknown *Contributors*: Hydrometeorological Prediction Center. Original uploader was Thegreatdr at en.wikipedia

Image:sec.gif *Source*: http://en.wikipedia.org/w/index.php?title=File:Sec.gif *License*: unknown *Contributors*: Epolk

file:US-StormPredictionCenter-Logo.svg *Source*: http://en.wikipedia.org/w/index.php?title=File:US-StormPredictionCenter-Logo.svg *License*: unknown *Contributors*: U.S. Government

Image:SPC severe outlook 04072006.png *Source*: http://en.wikipedia.org/w/index.php?title=File:SPC_severe_outlook_04072006.png *License*: unknown *Contributors*: Storm Prediction Center, NWS. Original uploader was Runningonbrains at en.wikipedia

File:Mesoscale discussion 2009-05-04.gif *Source*: http://en.wikipedia.org/w/index.php?title=File:Mesoscale_discussion_2009-05-04.gif *License*: unknown *Contributors*: Brynn Kerr, mesoscale forecaster at the Storm Prediction Center

File:2007 tornado watch 232.gif *Source*: http://en.wikipedia.org/w/index.php?title=File:2007_tornado_watch_232.gif *License*: unknown *Contributors*: Jack Hales, w:Storm Prediction CenterStorm Prediction Center

File:Day 1 fire outlook October 21, 2007.png *Source*: http://en.wikipedia.org/w/index.php?title=File:Day_1_fire_outlook_October_21,_2007.png *License*: unknown *Contributors*: Crosbie

Image:NHC TAFB.jpg *Source*: http://en.wikipedia.org/w/index.php?title=File:NHC_TAFB.jpg *License*: unknown *Contributors*: TAFB/NHC/NOAA

Image:CDC Example Graphic.gif *Source*: http://en.wikipedia.org/w/index.php?title=File:CDC_Example_Graphic.gif *License*: unknown *Contributors*: Monkeybait, Triberocker

Printed by Books on Demand GmbH, Norderstedt / Germany